Advances in Asian Human-Environmental Research

The *Advances in Asian Human-Environmental Research* series aims at fostering the discussion on the complex relationships between physical landscapes, natural resources, and their modification by human land use in various environments of Asia. It is widely acknowledged that human-environmental interactions become increasingly important in Area Studies and development research, taking into account regional differences as well as bio-physical, socio-economic and cultural particularities. The book series seeks to explore theoretic and conceptual reflection on dynamic human-environment systems applying advanced methodology and innovative research perspectives. The main themes of the series cover urban and rural landscapes in Asia. Examples include topics such as land and forest degradation, vulnerability and mitigation strategies, natural hazards and risk management concepts, environmental change, impact studies and consequences for local communities. The relevant themes of the series are mainly focused on geographical research perspectives of Area Studies, however there is scope for interdisciplinary contributions from various spheres within the natural sciences, social sciences and humanities. Key themes: Human-Environment Interaction - Asian regional studies - Asian geography - Impact studies - Landscape - Society - Land use and land cover change - Natural resources *Submit a proposal*: Proposals for the series will be considered by the Series Editor and International Editorial Board. An initial author/editor questionnaire and instructions for authors can be obtained from the Publisher, Dr. Robert K. Doe (robert.doe@springer.com).

Altaf Hussain

Urbanisation and Development Challenges in Kargil, Ladakh

The Transformation of the Indian Trans-Himalayan Perspective

Springer

Altaf Hussain
South Asia Institute
Universität Heidelberg
Heidelberg, Baden-Württemberg, Germany

ISSN 1879-7180 ISSN 1879-7199 (electronic)
Advances in Asian Human-Environmental Research
ISBN 978-3-032-05666-5 ISBN 978-3-032-05667-2 (eBook)
https://doi.org/10.1007/978-3-032-05667-2

This Springer imprint is published by the registered company Springer Nature Switzerland AG
The registered company address is: Gewerbestrasse 11, 6330 Cham, Switzerland

Preface

The Trans-Himalayan region, with its fragile ecosystems and rich cultural heritage, presents a unique opportunity to explore the evolving relationship between humans and their environment. Urbanisation in this region, particularly in Ladakh, has emerged as a transformative force, driving significant socio-economic and environmental changes. However, despite its growing importance, urbanisation in high-altitude regions remains an underexplored phenomenon. Ladakh, historically known for its remote mountain settlements, isolation, and tradition of self-reliance, has undergone profound transformations in recent years. Once defined by its rugged terrain and limited connectivity, the region is now experiencing sweeping changes driven by increased accessibility, administrative reforms, geopolitical shifts, rapid tourism growth, and government initiatives. These forces have collectively reshaped Ladakh's socio-economic and cultural fabric, bringing both opportunities and challenges.

This book focuses on Ladakh, particularly Kargil, to examine the urbanisation and development challenges in an ecologically fragile high-altitude environment. What makes this book unique is its grounded approach, combining empirical research place-based analysis, and first-hand experiences to offer a holistic understanding of Ladakh's urban transformation. It contributes to the growing discourse on urbanisation in high-altitude regions, providing valuable insights for researchers, urban planners, and policymakers. Through its analysis, the book not only deepens our understanding of Ladakh's evolving urban landscape but also offers broader lessons for similar Himalayan urban centres.

Ultimately, this book serves as a call to action, emphasising the critical need for thoughtful and inclusive planning to preserve Ladakh's fragile ecosystems and rich cultural heritage amid rapid transformation. How can we balance development with sustainability in such a unique and vulnerable region? This question lies at the heart of our exploration. Based on extensive research and field observations, this book seeks to inspire meaningful dialogue and action towards a more sustainable future for Ladakh and beyond.

Heidelberg, Germany Altaf Hussain

Acknowledgements

The book is the outcome of my doctoral research carried out at the Department of Geography, South Asia Institute, Heidelberg University. This would not have been possible without many people's constant support, guidance, and generosity. Firstly, I am deeply grateful to my supervisor, Prof. Dr. Marcus Nüsser, Department of Geography, South Asia Institute, Heidelberg University, for his guidance, encouragement, and support throughout the tenure of my doctoral journey. His mentorship has been instrumental in the successful completion of my doctoral degree. I want to express my deep gratitude to Prof. Dr. Rahul Mukherji for his expertise and co-supervision. In addition, I would like to thank the Ministry of Tribal Affairs, Government of India, for scholarship funding throughout my doctoral research. I am also thankful to the Consulate General of India, Munich, Germany, for their coordination and cooperation regarding official works and for providing scholarship funding. I am grateful to the Graduate Academy of Heidelberg University for granting the completion grant funded by the LGF (Landesgraduiertenförderung) programme during the crucial final phase of my doctoral degree. I extend my humble appreciation to former retired Prof. Dr. Atiya Habeeb Kidwai, Centre for the Study of Regional Development, Jawaharlal Nehru University, for her guidance, advice, and support particularly before coming to the Heidelberg University.

I further would like to express my gratitude to my colleagues Mohd Soheb, Dagmar Brombierstäudl, Javiera Machuca, Nicolas Zanetta, Kunzang Angmo, Atif Ishaq, and Huda Jawed for their insights and suggestions. I am grateful to former and senior colleagues Stanzin Passang, Maike Petersen, Judith Müller, Padma Ladon, and Isabell Haag. I also thank my friends Mohd Ibrahim, Abhimanyu Pandey, and Tsering Dorjay for their support, encouragement, and wisdom. I want to express my deep gratitude to Joanna Cichecka for her invaluable support in dealing with official progress reports and to Nils Harm for his assistance in cartography. I thank Dr. Susanne Schmidt, Prof. Matthias Weniger, Dr. Rakesh Bhambri, and Dr. Juliane Dame for their valuable feedback. Dr. Susanne Schmidt and Mohd Soheb offered their generous help in remote sensing, particularly in data acquisition for satellite images. Additionally, I sincerely thank Dr. Susanne Schmidt and Dagmar Brombierstäudl for their contributions to proofreading this book.

Data collection, including fieldwork, was conducted by visiting the Kargil district during the summer and autumn periods of 2021–2024 in Ladakh, India. Many people and friends supported and helped during the fieldwork in Kargil, Ladakh. I want to thank and appreciate my friends Shujat Ali Mir (Asst. Prof.), Iftikhar Hussain, Baqir Ali Rehmani, Irshad Hussain, Feroz Hussain, Dr. Altaf Hussain (MS Ortho), Zakir Hussain, and Mohd Hassanain for their company, kindness, and support during data collection. I sincerely appreciate all my friends who have supported me throughout my academic journey and contributed in various ways. I am grateful to all the members and alumni of All Kargil Students Association Delhi (AKSAD) for their guidance, support, and companionship during my academic journey in Delhi. I extend my sincere, heartfelt gratitude to all the participants for dedicating their time to participate in the interview. Your perspectives and insights were invaluable and significantly contributed to the depth and richness of my scientific research. The face-to-face talks, interviews, and discussions were conducted with local people, official persons, and stakeholders of the Kargil district in Ladakh. Being a resident of Kargil and knowing the language, I found it very easy to deal with people during data collection. I would like to extend my thanks to all the people who provided me with every type of support. I am thankful to the departments of LAHDC-Kargil in the UT of Ladakh for rendering me project reports, documents, and statistical data. I am also grateful to Monique Fort, Nicky Harman and Roger Bates, and the American Geographical Society Library at the University of Wisconsin-Milwaukee Libraries for their permission to reproduce the historical photographs.

I am writing to express profound gratitude to my family for their unwavering love, support, and strength throughout my journey. My parents, Ghulam Ahmad and Fatima Banoo, have been instrumental in shaping my path with their endless trust, reassurance, and moral support. This scientific research work is dedicated to the cherished memory of my grandfather, the late Haji Mohammad Ali, whose inspiration continues to guide me. The guidance and encouragement of my grandmother, Hajiya Kulsum Bee, and grandfather have profoundly influenced my academic journey. I am also deeply grateful to my elder brother, Mohd Abbas, for his constant moral support and motivation throughout my academic career. Heartfelt thanks to my sisters, Yasmeen Batool, Amina Banoo, and Leela Banoo, for their kindness and encouragement throughout my doctoral research efforts. I thank my wife, Asiya Banoo, for her love and support. Gratitude extends to my nephews and nieces, Meesum Abass, Wilayat Ali, Ali Asghar, Zehra Fatima, Zainab Fatima, and Fatima Kousar, for their affection and joy. I also appreciate the support and acknowledgement from my brothers-in-law, Gulzar Hussain and Mohd Hassan. Lastly, I extend heartfelt thanks to all relatives, the local community of Lànkore, and the people of Kargil in Ladakh for their immense moral support, encouragement, and blessings.

Heidelberg, Germany Altaf Hussain

Summary

Urbanisation has been one of the significant drivers of socio-economic and environmental change in the Trans-Himalaya. Recently, the remote and less accessible areas of Ladakh have undergone rapid urbanisation due to road construction, mountain tourism, geopolitical factors, administrative changes, and external interventions. These driving factors are responsible for the drastic transformation and development in Ladakh's newly formed Union Territory (UT). The region has experienced rapid urban growth and unplanned urbanisation, particularly in the capital towns of Leh and Kargil. After Leh, the capital, Kargil town is the second-largest urban centre in the region. Kargil town has witnessed significant changes in its urban landscape, driven by historical developments, urban population growth, rural-to-urban migration, spatial expansion, and the growing tourism sector. These factors have collectively contributed to the emergence, growth, and transformation of this mountain town in Ladakh. Based on a multi-temporal approach using high-resolution satellite images combined with a mixed-methods approach and field surveys, urbanisation and its landscape dynamics have been analysed since the 1960s.

The massive transformation in Ladakh began with the opening of the region to motorised traffic in 1960 and the introduction of tourism in 1974. These developments have led to significant changes in land use and land cover (LULC), as well as a shift in socio-economic activities across both rural and urban areas. The establishment of the Ladakh Autonomous Hill Development Council (LAHDC) in Leh in 1995 and in Kargil in 2003 has also provided political empowerment to the local population, enabling them to govern and make decisions regarding local planning, infrastructure, and development initiatives. In Kargil, the formation of LAHDC-Kargil in 2003 has spurred road construction, growth in tourism, improvements in education and health care, job creation, and infrastructure development. As the administrative centre, Kargil town attracts more people due to expanding opportunities in these sectors, leading to urban migration from rural areas. This influx has contributed to population growth and a surge in the construction of residential, commercial, and public buildings. In addition, the massive increase in tourist arrivals has led to the establishment of more hotels, guesthouses, and travel agencies.

Urban development schemes have played a vital role in shaping infrastructure and transforming the Kargil town. However, the lack of regulated planning has resulted in uncontrolled urban growth and haphazard expansion. Consequently, several critical urban challenges have emerged, including issues related to municipal solid waste management, water supply, public sanitation, deep drains, and urban mobility. These problems are primarily rooted in weak governance structures, unplanned development, and the absence of a coherent urban planning framework. As a consequence, the effective management and addressing of these complex urban issues pose a major challenge, highlighting the urgent need for integrated planning and robust governance.

Moreover, a significant transition has occurred in socio-economic activities, with the occupational structure shifting from agricultural to non-agricultural activities, reflecting the Ladakh's broader economic transformation. This study aims to contribute to the understanding of the urbanisation process in mountain areas, which still requires further investigation, and to provide the best possible recommendations for policymakers by critically examining emerging urban phenomena. The research suggests that greater emphasis should be placed on sustainable development to address urban planning and development challenges. It also highlights the importance of balancing modernisation with the preservation of cultural heritage and traditional practices to ensure holistic development. Adopting these recommendations can guide Ladakh towards an inclusive and sustainable future, fostering growth and development while preserving its unique identity.

Contents

About the Author

Altaf Hussain holds a Ph.D. from the Department of Geography at the South Asia Institute (SAI), Heidelberg University, Germany. His academic journey is marked by diverse and specialised studies. He received his master's degree in Planning with a specialisation in Regional Planning from the School of Planning and Architecture (SPA), New Delhi, India. Additionally, he earned his master's and bachelor's degree in Geography from the Department of Geography, Jamia Millia Islamia (JMI), New Delhi, India. He is also an associate member of the Institute of Town Planners, India (ITPI), reflecting his commitment to and recognised expertise in the field of urban and regional planning. His professional and academic pursuits are deeply rooted in the study of human geography, urban and regional planning, disaster management, sustainable development, and human-environment relationships. His research focuses on the mountainous regions of the Himalaya and Hindu Kush Himalaya, with a particular emphasis on the Trans-Himalayan region of Ladakh, India. In these regions, he explores the intricate dynamics between people and their environment, investigating how these relationships are shaped by and contribute to the distinct geographical and environmental conditions of high-altitude mountainous landscapes. The research further incorporates a critical examination of the planning and development dimensions of these Himalayan environments, particularly in relation to the complex challenges of managing growth, infrastructure, and sustainability in fragile mountain settings.

Abbreviations

AMRUT	Atal Mission for Rejuvenation and Urban Transformation
ASTER	Advanced Spaceborne Thermal Emission and Reflection Radiometer
BRO	Border Roads Organisation
CRIF	Central Road and Infrastructure Fund
DDMA	District Disaster Management Authority
DIP	District Irrigation Plan
FSTP	Faecal Sludge Treatment Plant
GIS	Geographic Information System
GLOFs	Glacial Lake Outburst Floods
H&UDD	Housing and Urban Development Department
HKH	Hindu Kush Himalaya
I&FC	Irrigation and Flood Control
IDMT	Integrated Development of Medium Towns
IDSMT	Integrated Development of Small and Medium Towns
IHSDP	Integrated Housing and Slum Development Programme
IKMT	Imam Khomeini Memorial Trust
ISK	Islamia School Kargil
JJM	Jal Jeevan Mission
JNNURM	Jawaharlal Nehru National Urban Renewal Mission
KDA	Kargil Democratic Alliance
KDA	Kargil Development Authority
KREDA	Kargil Renewable Energy Development Agency
LAB	Leh Apex Body
LAHDC	Ladakh Autonomous Hill Development Council
LAHDC-Kargil	Ladakh Autonomous Hill Development Council Kargil
LAHDC-Leh	Ladakh Autonomous Hill Development Council Leh
LBA	Ladakh Buddhist Association
LOC	Line of Control
LULC	Land Use and Land Cover
MC	Municipal Committee

MLC	Maximum Likelihood Classification
MoRTHs	Ministry of Road Transport and Highways
MoUD	Ministry of Urban Development
NAC	Notified Area Committee
NASA	National Aeronautics and Space Administration
NGO	Non-Governmental Organisation
NHIDCL	National Highways and Infrastructure Development Corporation Limited
PHE	Public Health Engineering
PMAY	Pradhan Mantri Awas Yojana
PMGSY	Pradhan Mantri Gram Sadak Yojana
PMKSY	Pradhan Mantri Krishi Sinchayee Yojana
PMRP	Prime Minister Reconstruction Programme
PWD	Public Works Department
R&B	Roads and Buildings
RAY	Rajiv Awas Yojana
SBM	Swachh Bharat Mission
SCM	Smart Cities Mission
SRTM	Shuttle Radar Topography Mission
TDA	Tourism Development Authority
UDPFI	Urban Development Plan Formulation and Implementation
UIDSSMT	Urban Infrastructure Development Scheme for Small and Medium Towns
ULBs	Urban Local Bodies
USGS	United States Geological Survey
UT	Union Territory
WDP	Watershed Development Programme

Chapter 1
Introduction

Abstract Urbanisation is a multifaceted socio-economic process that encompasses various dimensions and influences multiple aspects of human life and the environment. It typically involves the movement of people from rural to urban areas, leading to the expansion of towns and cities. This chapter examines the dynamics of mountain urbanisation in Ladakh, a region situated in the Trans-Himalayan region of India. It begins with a background review that highlights the urbanisation and development challenges in this high-altitude region. The study seeks to offer an in-depth insight into how urbanisation unfolds in Ladakh, considering key influencing factors such as socio-economic dynamics and policy interventions. Additionally, it explores the implications of urban growth on local communities, infrastructure, and environmental sustainability. Furthermore, the book outlines its specific objectives and research questions, establishing a clear framework for analysis. The structure of the book is also discussed, presenting a chapter-wise overview that ensures a systematic and in-depth examination of the subject.

Keywords Urbanisation · Development · Population · Socio-economic · Infrastructure · Challenges · Kargil · Ladakh · Trans-Himalayan region · India

1.1 The Himalayan Region

The global population has more than tripled since the mid-twentieth century, with over 55% now residing in urban areas (United Nations, 2019, 2022). In developing countries, 52% of the population lived in urban settings as of recent estimates, a figure projected to increase to 57% by 2030 and 66% by 2050 (Khor et al., 2022). South Asia alone witnessed an urban population increase of 130 million between 2000 and 2011—exceeding the entire population of Japan (Ellis & Roberts, 2016). This figure is expected to rise by nearly 250 million by 2030, with the urban population growing at an annual rate of over 2.5%, or approximately 14 million people (Ellis & Roberts, 2016). The region is heavily dependent on glacial water sources,

A. Hussain, *Urbanisation and Development Challenges in Kargil, Ladakh*, Advances in Asian Human-Environmental Research, https://doi.org/10.1007/978-3-032-05667-2_1

making it particularly vulnerable to climate change and disruptions in natural hydrological cycles (Mani, 2021). The Hindu Kush Himalayan (HKH) region is of vital importance for Asia, given its resources such as freshwater, carbon reserves, energy potential, human, agricultural, and biological diversity (Mukherji et al., 2018). It also delivers essential ecosystem services—such as water, food, and energy—that directly support the livelihoods of approximately 240 million people (Wester et al., 2019). The HKH region is undergoing rapid transformation, primarily driven by climate change and urbanisation, alongside contributing factors such as infrastructure development, globalisation, migration, and tourism (Mukherji et al., 2018; Wester et al., 2019). In addition, environmental and socio-economic changes in the region are being fuelled by LULC changes, the exploitation of natural resources, economic development, and increasing global interconnectedness (Banskota, 2000; Wang et al., 2019; Wester et al., 2019).

The Himalaya is a region of remarkable linguistic, cultural, and religious diversity, encompassing a vast and complex geographical expanse. Stretching from the borders of Myanmar and the Tibetan Plateau in China, across northern India, Bhutan, and Nepal, to Pakistan, it represents a physically immense and ecologically sensitive landscape (Chand, 2000; Tiwari, 2000). The region is geodynamically unstable and ranks among the most tectonically active mountain systems in the world, characterised by ongoing uplift and high vulnerability to natural hazards (Anees et al., 2022). Fragile mountain geological formations and ecosystems in the Himalaya are deteriorating rapidly due to both natural processes and anthropogenic pressures (Wester et al., 2019). In the Himalaya, urbanisation has been a significant driver of global environmental change particularly in transforming the small and large urban centres.

In 2015, the Indian Himalaya accounted for the largest share of the mountain population within the HKH region, with approximately 50.31 million people out of a total of 225 million residing in the region's mountainous and hilly areas (United Nations, 2015; Wester et al., 2019). This population concentration has been closely linked to significant environmental and socio-economic transformations, particularly through the expansion and intensification of urban land use (Bhandari, 2012; Tiwari et al., 2018; Wang et al., 2019). The Himalayan towns have encountered substantial challenges arising from ecological fragility, political shifts, and socio-economic changes (Dame et al., 2019; Gondhalekar et al., 2015; Kovács et al., 2019; Ojha et al., 2020). The rapid pace of urbanisation has exacerbated infrastructure deficits and environmental stresses (Haack & Rafter, 2006). Notably, capital cities in the Himalayan region experienced substantial urban expansion between 1972 and 2015, largely driven by a sharp rise in urban populations. For instance, Srinagar expanded from 9.4 to 142.2 km^2, Kathmandu from 11.4 to 92.6 km^2, and Dehradun from 4.1 to 50.1 km^2 (Diksha & Kumar, 2017). The urban centres in the Indian Himalayan region have experienced unplanned urban growth and rapid urbanisation, leading to several challenges, such as water crises (Bharti et al., 2020; Kovács et al., 2019; Müller, 2022; Sarkar, 2023), solid waste management (Dutta et al., 2020; Masood & Ahmad, 2020; Mushtaq et al., 2020), and built-up area changes (Dame et al., 2019; Hussain et al., 2023a; Maithani, 2020; Rashid & Aneaus, 2020). In conjunction with the impacts of climate change, the risk of cryospheric hazards and cloudburst-induced flooding is rising significantly (Bhatt et al., 2017; Mueller et al., 2019; Schmidt et al., 2020; Sekhri et al., 2020; Thayyen et al., 2013; Verma et al., 2022; Ziegler et al., 2016). Climate change is also disrupting the hydrological cycle by altering temperature regimes, precipitation patterns, and increasing the deposition of anthropogenic black carbon. These changes are accelerating snowmelt and glacier retreat across the Himalayan region (Mani, 2021).

1.2 The Trans-Himalayan Region

The Trans-Himalaya is a part of the lofty mountains of the Himalaya. The direction extends from south-east to north-west, and its extreme length is upwards of 563 km (Cunningham, 1854). On the north, Trans-Himalaya forms the natural boundary between Ladakh, Baltistan, and Rongdo; and Rupshu, Zanskar, Kargil (Purig[1]), Drass, and Astor on the south (Cunningham, 1854). Situated on the western end of the Himalayas, Ladakh passes through mountain ranges of the Great Himalaya, Ladakh, and Karakoram, and is geographically separated from the Indian subcontinent by the Great Himalayan Range and bordered to the north by the Karakoram Range (Dame & Nüsser, 2011; Sapru, 2001). Ladakh's ecosystem is highly sensitive and has been subject to climatic variability, unsustainable natural resource exploitation, and a range of natural hazards and disasters over the past several decades (Field, 2018; Schmidt et al., 2020; Soheb et al., 2022).

Historically, Ladakh was one of the crossroads of Trans-Himalayan and Trans-Karakoram trade between Tibet, India, Afghanistan, and Central Asia (Fewkes, 2008; Rizvi, 2012). It is a region with many diverse and often isolated rural groups (Raghuvanshi et al., 2017). Due to the region's harsh climatic conditions, extensive forest cover is absent; instead, the land-use pattern is characterised by the presence of willow and poplar trees, fruit orchards, and the collection of wild herbs (Dame & Nüsser, 2011). In recent decades, human activities and ecological processes have significantly reshaped the landscape of Ladakh (Barrett & Bosak, 2018). The introduction of tourism in 1974 marked a turning point for the region, stimulating economic growth in few towns and villages—most notably Leh and Kargil (Fox et al., 1994). Leh, the capital of Ladakh, has experienced substantial urban expansion driven by tourism, geopolitical significance, administrative functions, and infrastructure development (Dame et al., 2019). This urban growth has been accompanied by a transition from traditional agricultural livelihoods to income-generating opportunities primarily dependent on the tourism industry (Dame & Nüsser, 2011; Müller et al., 2020; Nüsser et al., 2012). Leh town has undergone unprecedented increase in hotels and buildings which have been built on agricultural land (Geneletti & Dawa, 2009;

[1] Kargil is mainly populated by a people known as Purigpa. The local name of the region had always been Purig, and it changed to Kargil (when this place became tehsil headquarter) particularly after the Dogra administration made the town their garrison headquarter in 1846 (Schettler & Schettler, 1989).

Goeury, 2010). At present, the tourism-driven economic transformation is reshaping livelihood patterns and demographic structures, resulting in increased pressure on natural resources across both rural and urban areas of Ladakh (Soheb et al., 2022). The major development projects like hydropower, road construction, airport, and other infrastructural projects are such examples, which have come from state and central government recently in Ladakh (Clouse, 2021).

A substantial number of military personnel are stationed in Ladakh due to its geopolitical significance and ongoing border tensions with Pakistan and China. This has led to extensive investments in road construction and public infrastructure across the Trans-Himalayan region (Aggarwal & Bhan, 2009; Baghel & Nüsser, 2015; Demenge, 2011). Alongside these developments, irrigated agriculture remains a vital component of livelihood security for mountain communities in the north-western Himalayan arc and its adjoining areas (Nüsser et al., 2019b). Irrigation water is harnessed for agricultural fields through runoff and reservoirs (rzing) using snow- and glacier-fed irrigation systems (Hill, 2014; Nüsser et al., 2012; Nüsser et al., 2019b). The inhabitants of the arid and semi-arid high-altitude regions are particularly reliant on irrigated agriculture to sustain their livelihoods (Kreutzmann, 2000). The government intervention road-building programmes have encouraged a shift towards greater dependence on subsidised goods and services in the communities of Ladakh (Dame & Mankelow, 2010). The livelihood and food security of people in Ladakh's permanent settlements depend on agriculture and animal husbandry (Dame & Nüsser, 2011; Dollfus, 2013).

The introduction of the Watershed Development Programme (WDP) in 1996 brought nearly thousands of hectares under irrigation, and the Ladakh Autonomous Hill Development Council Kargil (LAHDC-Kargil) undertook a number of smaller schemes for the development of water resources and the construction of irrigation infrastructure in Ladakh (Mankelow, 2003; Rizvi, 2012). The Pradhan Mantri Krishi Sinchayee Yojana (PMKSY) was one of the recent canal irrigation and infrastructure programmes introduced by LAHDC-Kargil in the district (Hussain et al., 2023a). After the Kargil War 1999, the Indian army initiated several civil society programmes like Operation Sadbhavana (Goodwill mission) to provide health and education services in the borderland conflict areas (Aggarwal & Bhan, 2009; Bhan, 2013). From 1995, Kargil town was supplied with electricity from the 3.75 MW Iqbal hydel project and by the end of 2008–09, the number of diesel-powered powerhouses had increased to 34 (Census of India, 2011a; Rizvi, 2012). In 2013, the 44-megawatt Chutuk hydroelectric project was constructed on the Suru River in Chutuk village. This project now supplies electricity to both rural and urban populations in the Kargil district. To promote and develop renewable energy, the Kargil Renewable Energy Development Agency (KREDA), a Non-Governmental Organisation (NGO) based in Kargil, is working on solar, wind, and hydro energy projects. KREDA has recently implemented several micro-hydro projects and solar power initiatives to ensure sustainable energy solutions in the remote areas of Kargil district in Ladakh.

In recent years, several studies have addressed the socio-economic and environmental challenges in Ladakh, particularly focusing on mountain urbanisation, pastoralism, food security, socio-hydrological systems, glacier changes, and borderland issues. Research on mountain urbanisation (Dame et al., 2019; Hussain et al., 2023a; Müller & Dame, 2016) explores the issues stemming from rapid urban growth, such as infrastructure deficits, migration patterns, housing, and the pressure on natural resources in high-altitude areas. At the same time, traditional livelihoods like pastoralism, which have been central to food security, are facing challenges due to climate change, changing grazing patterns, and loss of pastureland (Dame, 2018; Dame & Nüsser, 2011; Dollfus, 2013; Ladon et al., 2023). The region's socio-hydrological systems, which are heavily dependent on glaciers and snowmelt, are also under pressure as water systems are increasingly impacted by climate change and evolving agricultural practices (Müller, 2022; Nüsser et al., 2012). Research on glacier changes (Nüsser et al., 2019a; Schmidt & Nüsser, 2012; Schmidt et al., 2020; Soheb et al., 2020) reveals the significant retreat of glaciers, which has wide-ranging effects on water availability and the landscape. Studies on the cryosphere (Brombierstäudl et al., 2021, 2022; Nüsser et al., 2019b, 2022; Passang et al., 2022; Schmidt et al., 2020) further explore the accelerated melting of glaciers and the consequences for water security and local livelihoods. Additionally, the region's geopolitical situation has drawn attention to borderland conflicts and their impact on tourism and local communities, with studies (Aggarwal, 2004; Bhan, 2013; Geneletti & Dawa, 2009; Gupta, 2022; Pelliciardi, 2021) examining how these conflicts shape economic activities, exacerbate tensions, and influence the development of tourism.

1.3 Dynamics of Urbanisation

Urbanisation is a multifaceted socio-economic process that reshapes the built environment and drives a demographic shift from rural to urban settlements (United Nations, 2019). It is significantly influenced by the scale and nature of social, economic, and political transformations (Tacoli, 2017). According to the Census of India, an area is classified as urban if it has a minimum population of 5,000, a population density of no less than 400 persons per km^2, at least 75% of the male main workforce engaged in non-agricultural occupations, and is officially designated as a statutory town—such as a municipality, corporation, cantonment board, and notified town area committee (Census of India, 2011b). In India, urbanisation has emerged as a critical and irreversible trend, playing a vital role in national economic development and poverty reduction (Ministry of Housing & Urban Affairs, 2020). The country has witnessed a substantial rise in urban population over recent decades. The urban share of the population increased from 23% in 1980 to 32% in 2011, and reached 34% by 2018, while the total population grew more than threefold since 1950, reaching 1.35 billion (United Nations, 2019; World Bank, 2013). According to a World Bank estimate, India will require approximately \$840 billion in capital investment for urban infrastructure and municipal services over the next 15 years to meet the demands of its rapidly urbanising population (Athar et al., 2022). Cities are the engines of economic growth. The urban sector's contribution to India's GDP rose from 29% in 1950–51 to 47% in 1980–81, reaching approximately 62–63% by 2008 (Planning Commission, 2008). Projections suggest that this figure may rise to about 75% by 2030 (Ministry of Urban Development, 2015; Planning Commission, 2008). Concurrently, the number of towns increased significantly—from 3,768 in 1991 to 5,161 in 2001, and further to 7,935 in 2011—resulting in a net addition of 2,774 towns (comprising 2,532 census towns and 242 statutory towns) between the 2001 and 2011 censuses (Bhagat, 2011; Planning Commission, 2008).

Urbanisation has become a major driver of environmental change in the transformation of mountain regions (Tiwari et al., 2018). In the mountain regions in Chitral and Nanga Parbat, the evolution of the cultural landscape is characterised by the expansion of human settlements and corresponding extensions of cultivated land (Nüsser, 2000, 2001). The mountain region of Ladakh is experiencing rapid urbanisation, particularly in the administrative centres of Leh and Kargil. It is argued that urban development in Leh town involves the active participation of various stakeholders in restructuring and beautification efforts (Müller & Dame, 2016). Similarly, major driving factors such as historical development, urban population growth, rural-to-urban migration, spatial expansion, and tourism have played a significant role in shaping and transforming Kargil's urban landscape (Hussain et al., 2023a, 2023b). In Kargil town, the unplanned urban development and encroachment have led to increased critical urban challenges, including solid waste management, traffic congestion, sanitation issues, and drainage problems (Hussain et al., 2023a). In recent times, the town has witnessed increased rural-to-urban migration, along with an influx of domestic tourists, seasonal labourers, and business-related individuals from outside the region. These external factors are increasing the burden on urban services such as sanitation, solid waste management, water supply, public toilets, deep drains, and other essential infrastructure in this mountain town.

The urbanisation process has significantly altered the landscape character of small towns in Ladakh. The most notable transformations have occurred particularly after the opening of road connectivity in 1960, the introduction of tourism sector in 1974, and granting of LAHDC in Leh and Kargil. The formation of LAHDC-Kargil in 2003 triggered a rise in construction activities, an increase in rural-to-urban migration towards the administrative centre, and the development of road infrastructure. The rapid growth of the urban population, along with rural-to-urban migration, has led to numerous challenges, including traffic congestion, water supply shortages, solid waste management issues, high building densities, and difficulties in providing adequate basic urban services. Additionally, the influx of seasonal migrants during the summer months exacerbates the strain on Kargil's already inadequate infrastructure, as these migrants participate in various economic activities, such as tourism, agriculture, and construction. The tourism sector has grown significantly over the last two decades, particularly after the 1999 Kargil War. This growth has resulted in a sharp increase in tourist arrivals and the construction of tourism-related facilities. Furthermore, the development of road infrastructure has accelerated rapidly due to geopolitical circumstances. Subsistence agriculture, once the backbone of the local economy, is now under threat due to water scarcity caused by the shrinking and melting of glaciers. This has led to a shift towards non-agricultural activities, especially in small towns. Arable land that was traditionally used for farming is increasingly being repurposed for residential, commercial, and tourist infrastructure without proper land-use planning. For instance, in Leh, fertile areas such as Skara and Sankar (Changspa), which are popular among tourists, have been transformed to accommodate tourist infrastructure (Rieger-Jandl, 2005).

In Ladakh, settlements are primarily situated in valleys and along riverbanks to make use of the scarce arable land and available water resources. For example, Kargil town is developed on both sides of the Suru River, with new settlements emerging in the surrounding areas, especially on the Kurbathang Plateau. However, unplanned and uncontrolled construction practices in all urban settlements have led to haphazard growth. The absence of town planning has resulted in a lack of norms, regulations, and bylaws to govern construction activities. Furthermore, Ladakh has experienced a shortage of physical and social infrastructure. There is a need to develop adequate physical infrastructure, such as roads, public transport, parking spaces, clean drinking water, solid waste management, and a reliable electricity supply in both rural and urban areas. Additionally, there is a need to strengthen social infrastructure by providing schools, hospitals, playgrounds, libraries, and public spaces. To address these challenges, a robust long-term vision is essential for the overall urban and regional development of Ladakh. This vision should focus on sustainable growth, proper land-use planning, and the equitable distribution of resources to ensure a balanced and resilient future for the region.

1.4 Objectives and Research Questions

Urban planning in mountain areas has been largely neglected by bureaucrats and policymakers, particularly in the Trans-Himalayan region. This region is environmentally and culturally diverse, making many of the Indian government's norms, strategies, and policies difficult to apply and implement in this ecologically sensitive area. Haphazard and unplanned development has emerged as a major challenge, hindering efforts to address critical urban issues. Therefore, the establishment of a robust urban and regional planning framework is essential for achieving sustainable development in mountain regions. This research examines mountain urbanisation and development challenges in the context of Ladakh's Trans-Himalayan region. This study aims to provide a comprehensive understanding of urbanisation dynamics and assist in promoting sustainable development in mountainous regions. The case study focuses on Kargil town, the administrative capital of the Kargil district, along with the emerging urbanised settlements that have also been considered. The objectives of this study are to comprehensively analyse the underlying causes and effects of mountain urbanisation and the associated development challenges. The following objectives will be addressed:

1. To analyse the role of historical development in shaping the contemporary urban landscape dynamics.
2. To identify and assess the key driving factors influencing urban landscape dynamics with a particular focus on Kargil town.
3. To examine the major urban infrastructure challenges and their implications for urban development in Kargil town.
4. To investigate the underlying causes of LULC changes.
5. To formulate recommendations for promoting sustainable development based on empirical research findings.

The research questions are as follows:

a. What is the role of historical development?
b. What are the key driving factors of urban landscape dynamics?
c. What are the major prevailing urban infrastructure challenges?
d. What are the primary causes of LULC changes?
e. What are the best possible recommendations for promoting sustainable development?

1.5 Structure of the Book

This chapter deals with the introduction, including a background review, and offers an overview and detailed understanding of the urbanisation processes in the context of Ladakh, located in the Indian Trans-Himalaya. The objectives and research questions are also discussed in this chapter.

Chapter 2 focuses on geographical setting of the study area, covering its climate, culture, and economy. It highlights the region's physical features, including location, altitude, and topography, and explores its cultural heritage, traditions, languages, and religious practices. This chapter also examines key economic activities and their impact on the socio-economic landscape.

Chapter 3 presents the data collection and the methodological framework applied to the study area. To interpret and analyse the research area, the methodological approach of mixed methods combined with multi-temporal and multi-scale remote sensing is employed. This approach allows for a comprehensive analysis by integrating both qualitative and quantitative data, while utilizing remote sensing tools to capture changes over time and across different spatial scales. It ensures a thorough understanding of the region's dynamics and provides a robust foundation for the study's findings.

Chapter 4 describes the historical context of Ladakh's development, beginning with the collapse of the Tibetan monarchy in 842 A.D., followed by the Dogra rule, the Colonial period, the Trans-Karakoram trade, and the post-independence period. The chapter examines the region's political, cultural, and economic evolution through these periods, highlighting key events and shifts that shaped Ladakh's identity and development. It also explores the integration of Ladakh into the broader political frameworks of India and the changes brought by the Indian government, especially in terms of infrastructure development, political representation, and the challenges faced by Ladakh in the context of regional geopolitics.

Chapter 5 examines the dynamics of Kargil's urban landscape and its transformation, focusing on historical evolution, demographic shifts, urban expansion, built-up area development, and the growing tourism sector. It analyses the influence of the mountainous geography on infrastructure, building patterns, and urban structure, while also considering key socio-economic drivers such as migration, economic opportunities, and tourism growth. By exploring these factors, the chapter provides a comprehensive understanding of Kargil's urban landscape dynamics (Fig. 1.1).

Chapter 6 presents urban infrastructure challenges and their implications, offering a detailed analysis of town planning and urban infrastructure schemes in the context of the current urban situation. The chapter examines urban development that has

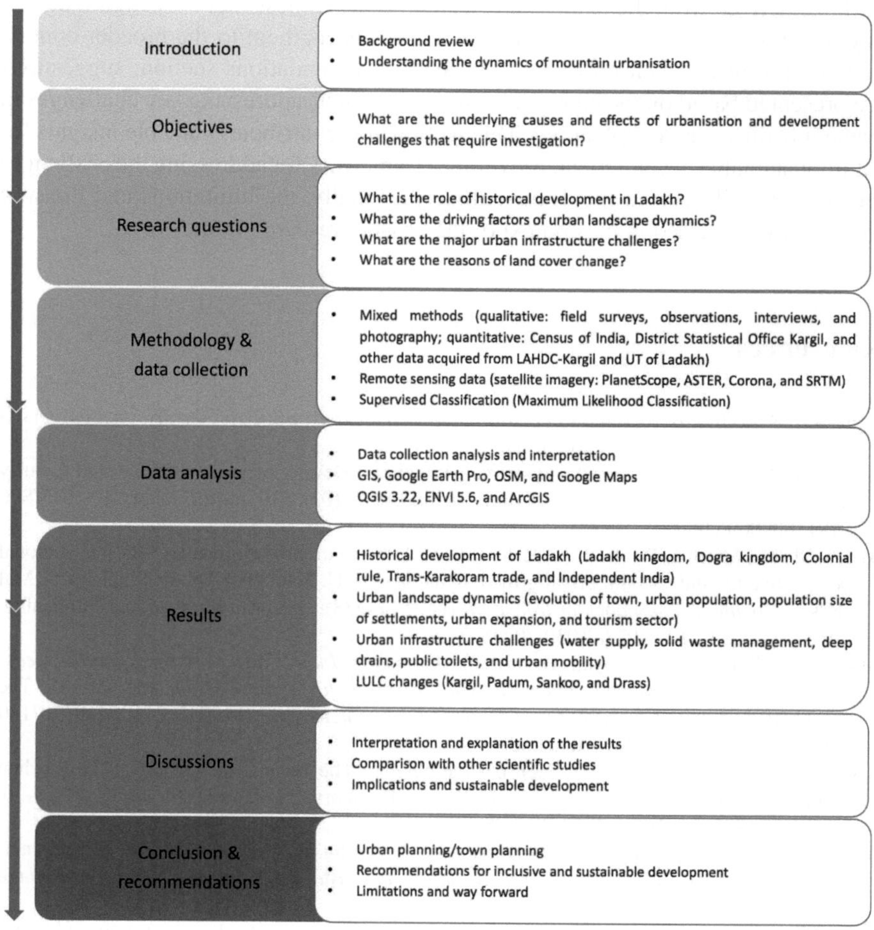

Fig. 1.1 The research framework of the study area in Ladakh. Draft by Altaf Hussain

reshaped the town's structure, highlighting critical urban challenges such as solid waste management, water supply, public toilets, drainage, and urban mobility. These issues arise from unplanned growth, inadequate governance structures, and the lack of comprehensive urban planning.

Chapter 7 focuses on LULC changes since the 1960s. The introduction of motorised traffic in 1960 and international tourism in 1974 have been key drivers of socio-economic transformation in Kargil, Ladakh. These developments led to an increase in external interventions, which have shifted agricultural patterns and economic activities, and have altered the landscape structure, particularly in Kargil town and urbanised settlements.

Chapter 8 deals with the discussion, conclusion, and recommendations of the study area. The discussion section analyses the empirical research and compares the results in context of other scientific studies in Ladakh as well as in the Trans-Himalayan region. The conclusion section addresses the research questions by synthesising the findings derived from the empirical research. This synthesis not only highlights the key insights derived from the study but also connects them to the broader context of the relevant research on Ladakh. In the recommendations section, suggestions are presented based on the empirical research findings, addressing key challenges in urban infrastructure and planning. This section also contributes valuable insights for future sustainable development, providing a framework for addressing the challenges faced by Kargil town in the Trans-Himalaya. Finally, the limitations and the way forward are discussed in the context of the broader research framework.

References

Aggarwal, R. (2004). *Beyond lines of control performance and politics on the disputed borders of Ladakh, India*. Duke University Press.

Aggarwal, R., & Bhan, M. (2009). "Disarming Violence": Development, democracy, and security on the borders of India. *The Journal of Asian Studies, 68*(02), 519. https://doi.org/10.1017/S0021911809000692

Anees, M. M., Sharma, R., & Joshi, P. K. (2022). Urbanization in Himalaya—An interregional perspective to land use and urban growth dynamics. In U. Schickhoff, R. B. Singh, & S. Mal (Eds.), *Mountain landscapes in transition* (pp. 517–538). Springer International Publishing. https://doi.org/10.1007/978-3-030-70238-0_23

Athar, S., White, R., & Goyal, H. (2022). *Financing India's urban infrastructure needs: Constraints to commercial financing and prospects for policy action* (pp. 1–78). The World Bank. https://documents1.worldbank.org/curated/en/099615110042225105/pdf/P17130200d91fc0da0ac610a1e3e1a664d4.pdf

Baghel, R., & Nüsser, M. (2015). Securing the heights: The vertical dimension of the Siachen conflict between India and Pakistan in the Eastern Karakoram. *Political Geography, 48*, 24–36. https://doi.org/10.1016/j.polgeo.2015.05.001

Banskota, M. (2000). The Hindu Kush-Himalayas: Searching for viable socioeconomic and environmental options. In *Growth, poverty alleviation and sustainable resource management in the mountain areas of South Asia* (pp. 57–105).

Barrett, K., & Bosak, K. (2018). The role of place in adapting to climate change: A case study from Ladakh, Western Himalayas. *Sustainability, 10*(4), 898. https://doi.org/10.3390/su10040898

Bhagat, R. B. (2011). Emerging pattern of urbanisation in India. *Economic and Political Weekly, 46*(34), 10–12. https://www.jstor.org/stable/23017782

Bhan, M. (2013). *Counterinsurgency, democracy, and the politics of identity in India.* Routledge. https://doi.org/10.4324/9781315889863

Bhandari, M. (2012). International centre for integrated mountain development. In G. Ritzer (Ed.), *The Wiley-Blackwell encyclopedia of globalization* (p. wbeog308). John Wiley & Sons, Ltd. https://doi.org/10.1002/9780470670590.wbeog308

Bharti, N., Khandekar, N., Sengupta, P., Bhadwal, S., & Kochhar, I. (2020). Dynamics of urban water supply management of two Himalayan towns in India. *Water Policy, 22*(S1), 65–89. https://doi.org/10.2166/wp.2019.203

Bhatt, C. M., Rao, G. S., Farooq, M., Manjusree, P., Shukla, A., Sharma, S. V. S. P., Kulkarni, S. S., Begum, A., Bhanumurthy, V., Diwakar, P. G., & Dadhwal, V. K. (2017). Satellite-based assessment of the catastrophic Jhelum floods of September 2014, Jammu & Kashmir, India. *Geomatics, Natural Hazards and Risk, 8*(2), 309–327. https://doi.org/10.1080/19475705.2016.1218943

Brombierstäudl, D., Schmidt, S., & Nüsser, M. (2021). Distribution and relevance of aufeis (icing) in the Upper Indus Basin. *Science of the Total Environment, 780*, 146604. https://doi.org/10.1016/j.scitotenv.2021.146604

Brombierstäudl, D., Schmidt, S., & Nüsser, M. (2022). Spatial and temporal dynamics of aufeis in the Tso Moriri basin, eastern Ladakh, India. *Permafrost and Periglacial Processes*, 2173. https://doi.org/10.1002/ppp.2173

Census of India. (2011a). *District census handbook Kargil: Village and town directory, series-02, part XII-A.* Directorate of Census Operations, Jammu & Kashmir.

Census of India. (2011b). *Urban agglomerations and cities.* https://censusindia.gov.in/2011-prov-results/paper2/data_files/india2/1.%20data%20highlight.pdf

Chand, R. (2000). Agricultural development, growth and poverty in India's mountain region. In M. Banskota, T. S. Papola, & J. Richter (Eds.), *Proceedings of the international conference held from 31 January–4 February 2000 in Kathmandu, Nepal* (pp. 275–291). International Centre for Integrated Mountain Development and German Foundation for International Development.

Clouse, C. (2021). *Climate-adaptive design in high mountain villages: Ladakh in transition.* Routledge, Taylor & Francis Group.

Cunningham, A. (1854). *Ladák, physical, statistical, and historical: With notes of the surrounding countries.* Wm. H. Allen and Company.

Dame, J. (2018). Food security and translocal livelihoods in high mountains: Evidence from Ladakh, India. *Mountain Research and Development, 38*(4), 310. https://doi.org/10.1659/MRD-JOURNAL-D-18-00026.1

Dame, J., & Mankelow, J. S. (2010). Stongde revisited: Land-use change in Central Zangskar. *Erdkunde, 64*(4), 355–370. https://doi.org/10.3112/erdkunde.2010.04.05

Dame, J., & Nüsser, M. (2011). Food security in high mountain regions: Agricultural production and the impact of food subsidies in Ladakh, Northern India. *Food Security, 3*(2), 179–194. https://doi.org/10.1007/s12571-011-0127-2

Dame, J., Schmidt, S., Müller, J., & Nüsser, M. (2019). Urbanisation and socio-ecological challenges in high mountain towns: Insights from Leh (Ladakh), India. *Landscape and Urban Planning, 189*, 189–199. https://doi.org/10.1016/j.landurbplan.2019.04.017

Demenge, J. (2011). *The political ecology of road construction in Ladakh.* University of Sussex. http://sro.sussex.ac.uk/38501/

Diksha, & Kumar, A. (2017). Analysing urban sprawl and land consumption patterns in major capital cities in the Himalayan region using geoinformatics. *Applied Geography, 89*, 112–123. https://doi.org/10.1016/j.apgeog.2017.10.010

Dollfus, P. (2013). Transformation processes in Nomadic Pastoralism in Ladakh. *Himalaya, the Journal of the Association for Nepal and Himalayan Studies, 32*(1), 15. http://digitalcommons.macalester.edu/himalaya/vol32/iss1/15

Dutta, N., Ghosh, A., Debnath, B., & Ghosh, S. K. (2020). Climate change in hilly regions of India: Issues and challenges in waste management. In S. K. Ghosh (Ed.), *Sustainable waste management: Policies and case studies* (pp. 657–669). Springer Singapore. https://doi.org/10.1007/978-981-13-7071-7_59

Ellis, P., & Roberts, M. (2016). *Leveraging urbanization in South Asia: Managing spatial transformation for prosperity and livability*. World Bank. https://doi.org/10.1596/978-1-4648-0662-9

Fewkes, J. H. (2008). *Trade and contemporary society along the silk road*. Routledge. https://doi.org/10.4324/9780203273692

Field, J. (2018). The impact on disaster governance of the intersection of environmental hazards, border conflict and disaster responses in Ladakh, India. *International Journal of Disaster Risk Reduction*, 9.

Fox, J. L., Nurbu, C., Bhatt, S., & Chandola, A. (1994). Wildlife conservation and land-use changes in the Transhimalayan region of Ladakh, India. *Mountain Research and Development, 14*(1), 39. https://doi.org/10.2307/3673737

Geneletti, D., & Dawa, D. (2009). Environmental impact assessment of mountain tourism in developing regions: A study in Ladakh, Indian Himalaya. *Environmental Impact Assessment Review, 29*(4), 229–242. https://doi.org/10.1016/j.eiar.2009.01.003

Goeury, D. (2010). Ladakh, kingdom of sustainable development?: Protecting the natural environment to protect identity. *Revue de Géographie Alpine, 98-1*. https://doi.org/10.4000/rga.1147

Gondhalekar, D., Nussbaum, S., Akhtar, A., & Kebschull, J. (2015). Planning under uncertainty: Climate change, water scarcity and health issues in Leh Town, Ladakh, India. In W. Leal Filho, & V. Sümer (Eds.), *Sustainable water use and management* (pp. 293–312). Springer International Publishing. https://doi.org/10.1007/978-3-319-12394-3_16

Gupta, R. (2022). *Freedom in captivity: Negotiations of belonging along Kashmir's frontier*. Cambridge University Press.

Haack, B. N., & Rafter, A. (2006). Urban growth analysis and modeling in the Kathmandu Valley, Nepal. *Habitat International, 30*(4), 1056–1065. https://doi.org/10.1016/j.habitatint.2005.12.001

Hill, J. (2014). Farmer-managed irrigation systems in Baltistan and Kargil. *Ladakh Studies, 31*, 4–23.

Hussain, A., Schmidt, S., & Nüsser, M. (2023a). Dynamics of mountain urbanisation: Evidence from the Trans-Himalayan town of Kargil, Ladakh, India. *Land, 12*(4), 920. https://doi.org/10.3390/land12040920

Hussain, A., Schmidt, S., & Nüsser, M. (2023b). *Urban landscape change in the Trans-Himalayan Town of Kargil, Ladakh, India*. https://doi.org/10.5194/egusphere-egu23-2663

Khor, N., Arimah, B., Otieno, R. O., van Oostrum, M., Mutinda, M., & Martins, J. O. (2022). *World cities report 2022: Envisaging the future of cities*. United Nations Human Settlements Programme (UN-Habitat).

Kovács, E. K., Ojha, H., Neupane, K. R., Niven, T., Agarwal, C., Chauhan, D., Dahal, N., Devkota, K., Guleria, V., Joshi, T., Michael, N. K., Pandey, A., Singh, N., Singh, V., Thadani, R., & Vira, B. (2019). A political ecology of water and small-town urbanisation across the lower Himalayas. *Geoforum, 107*, 88–98. https://doi.org/10.1016/j.geoforum.2019.10.008

Kreutzmann, H. (Ed.). (2000). *Sharing water: Irrigation and water management in the Hindukush, Karakoram, Himalaya*. Oxford University Press.

Ladon, P., Nüsser, M., & Garkoti, S. C. (2023). Mountain agropastoralism: Traditional practices, institutions and pressures in the Indian Trans-Himalaya of Ladakh. *Pastoralism, 13*(1), 30. https://doi.org/10.1186/s13570-023-00289-1

Maithani, S. (2020). A quantitative spatial model of urban sprawl and its application to Dehradun urban agglomeration, India. *Journal of the Indian Society of Remote Sensing, 48*(11), 1583–1592. https://doi.org/10.1007/s12524-020-01182-9

Mani, M. (Ed.). (2021). *Glaciers of the Himalayas: Climate change, black carbon, and regional resilience*. The World Bank. https://doi.org/10.1596/978-1-4648-0099-3

Mankelow, J. S. (2003). *The implementation of the watershed development programme in Zangskar, Ladakh: Irrigation development, politics and society*. University of London. https://www.soas.ac.uk/water/publications/papers/file38409.pdf

Masood, A., & Ahmad, K. (2020). Assessment of municipal solid waste management in Jammu City: Problems, prospects and solutions. In S. Ahmed, S. M. Abbas, & H. Zia (Eds.), *Smart cities—Opportunities and challenges* (Vol. 58, pp. 257–275). Springer Singapore. https://doi.org/10.1007/978-981-15-2545-2_23

Ministry of Housing and Urban Affairs. (2020). *Annual report 2020–21*. Government of India. https://mohua.gov.in/upload/uploadfiles/files/Annual_Report_2020_21_MoHUA_EnglishVersion%20(Final).pdf

Ministry of Urban Development. (2015). *Smart cities: Mission statement and guidelines*. Government of India. https://smartcities.gov.in/themes/habikon/files/SmartCityGuidelines.pdf

Mueller, S., Sammonds, P., Bhat, G. M., Pandita, S., Suri, K., Thusu, B., & Le Masson, V. (2019). Disaster scenario simulation of the 2010 cloudburst in Leh, Ladakh, India. *International Journal of Disaster Risk Reduction, 33*, 485–494. https://doi.org/10.1016/j.ijdrr.2018.09.004

Mukherji, A., Scott, C., Molden, D., & Maharjan, A. (2018). Megatrends in Hindu Kush Himalaya: Climate change, urbanisation and migration and their implications for water, energy and food. In A. K. Biswas, C. Tortajada, & P. Rohner (Eds.), *Assessing global water megatrends* (pp. 125–146). Springer Singapore. https://doi.org/10.1007/978-981-10-6695-5_8

Müller, J. (2022). *Urban mountain waterscapes in Leh, Indian Trans-Himalaya: The transformation of hydro-social relations*. Springer International Publishing. https://doi.org/10.1007/978-3-031-18249-5

Müller, J., & Dame, J. (2016). Small town, great expectations: Urbanization and beautification in Leh. *South Asia Multidisciplinary Academic Journal, 14*. https://doi.org/10.4000/samaj.4208

Müller, J., Dame, J., & Nüsser, M. (2020). Urban mountain waterscapes: The transformation of hydro-social relations in the trans-Himalayan town Leh, Ladakh, India. *Water, 12*(6), 1698. https://doi.org/10.3390/w12061698

Mushtaq, J., Dar, A. Q., & Ahsan, N. (2020). Spatial–temporal variations and forecasting analysis of municipal solid waste in the mountainous city of north-western Himalayas. *SN Applied Sciences, 2*(7), 1161. https://doi.org/10.1007/s42452-020-2975-x

Nüsser, M. (2000). Change and persistence: Contemporary landscape transformation in the Nanga Parbat Region, Northern Pakistan. *Mountain Research and Development, 20*(4), 348–355. https://doi.org/10.1659/0276-4741(2000)020[0348:CAPCLT]2.0.CO;2

Nüsser, M. (2001). Understanding cultural landscape transformation: A re-photographic survey in Chitral, eastern Hindukush, Pakistan. *Landscape and Urban Planning, 57*(3–4), 241–255. https://doi.org/10.1016/S0169-2046(01)00207-9

Nüsser, M., Dame, J., Kraus, B., Baghel, R., & Schmidt, S. (2019a). Socio-hydrology of "artificial glaciers" in Ladakh, India: Assessing adaptive strategies in a changing cryosphere. *Regional Environmental Change, 19*(5), 1327–1337. https://doi.org/10.1007/s10113-018-1372-0

Nüsser, M., Dame, J., Parveen, S., Kraus, B., Baghel, R., & Schmidt, S. (2019b). Cryosphere-fed irrigation networks in the Northwestern Himalaya: Precarious livelihoods and adaptation strategies under the impact of climate change. *Mountain Research and Development, 39*(2). https://doi.org/10.1659/MRD-JOURNAL-D-18-00072.1

Nüsser, M., Schmidt, S., & Dame, J. (2012). Irrigation and development in the Upper Indus Basin: Characteristics and recent changes of a socio-hydrological system in Central Ladakh, India. *Mountain Research and Development, 32*(1), 51–61. https://doi.org/10.1659/MRD-JOURNAL-D-11-00091.1

Nüsser, M., Schmidt, S., & Dame, J. (2022). *Cryosphere changes and local adaptation strategies: Socio-hydrological case studies from the Himalayan region*. https://doi.org/10.5194/iahs2022-661

Ojha, H., Neupane, K. R., Pandey, C. L., Singh, V., Bajracharya, R., & Dahal, N. (2020). Scarcity amidst plenty: Lower Himalayan cities struggling for water security. *Water, 12*(2), 567. https://doi.org/10.3390/w12020567

Passang, S., Schmidt, S., & Nüsser, M. (2022). Topographical impact on snow cover distribution in the Trans-Himalayan region of Ladakh, India. *Geosciences, 12*(8), 311. https://doi.org/10.3390/geosciences12080311

Pelliciardi, V. (2021). Factors affecting international and national tourist arrivals (1974–2020) in Leh District (UT Ladakh, India). *European Journal of Sustainable Development, 10*(1), 736–736. https://doi.org/10.14207/ejsd.2021.v10n1p736

Planning Commission. (2008). *Eleventh five year plan (2007–2012): Agriculture, rural development, industry, services and physical infrastructure* (Vol. III). Oxford University Press.

Raghuvanshi, M. S., Tewari, J. C., Dolma, R., Arunachalam, A., & Yadav, O. P. (2017). Struggle from subsistence to sustainability and threat to local biodiversity under changing climate: A case study on Ladakh folk agriculture. *Climate Change and Environmental Sustainability, 5*(1), 59. https://doi.org/10.5958/2320-642X.2017.00006.0

Rashid, I., & Aneaus, S. (2020). Landscape transformation of an urban wetland in Kashmir Himalaya, India using high-resolution remote sensing data, geospatial modeling, and ground observations over the last 5 decades (1965–2018). *Environmental Monitoring and Assessment, 192*(10), 635. https://doi.org/10.1007/s10661-020-08597-4

Rieger-Jandl, A. (2005). *Living culture in the Himalayas: Anthropological guidelines for building in developing countries.* WUV University Press, University of Vienna.

Rizvi, J. (2012). *Ladakh: Crossroads of high Asia* (3rd ed.). Oxford University Press.

Sapru, C. S. (2001). *District census handbook, Leh district: Village and town directory, series-2, part-A and B.* Deputy Director of Census Operations, Jammu & Kashmir.

Sarkar, S. (2023). Urban water crisis and the promise of infrastructure: A case study of Shimla, India. *Frontiers in Water, 5*, 1051336. https://doi.org/10.3389/frwa.2023.1051336

Schettler, M., & Schettler, R. (1989). *Kashmir, Ladakh & Zanskar: A travel survival kit* (3rd ed.). Lonely Planet.

Schmidt, S., & Nüsser, M. (2012). Changes of high altitude glaciers from 1969 to 2010 in the Trans-Himalayan Kang Yatze Massif, Ladakh, Northwest India. *Arctic, Antarctic, and Alpine Research, 44*(1), 107–121. https://doi.org/10.1657/1938-4246-44.1.107

Schmidt, S., Nüsser, M., Baghel, R., & Dame, J. (2020). Cryosphere hazards in Ladakh: The 2014 Gya glacial lake outburst flood and its implications for risk assessment. *Natural Hazards, 104*(3), 2071–2095. https://doi.org/10.1007/s11069-020-04262-8

Sekhri, S., Kumar, P., Fürst, C., & Pandey, R. (2020). Mountain specific multi-hazard risk management framework (MSMRMF): Assessment and mitigation of multi-hazard and climate change risk in the Indian Himalayan Region. *Ecological Indicators, 118*, 106700. https://doi.org/10.1016/j.ecolind.2020.106700

Soheb, M., Ramanathan, A., Angchuk, T., Mandal, A., Kumar, N., & Lotus, S. (2020). Mass-balance observation, reconstruction and sensitivity of Stok glacier, Ladakh region, India, between 1978 and 2019. *Journal of Glaciology, 66*(258), 627–642. https://doi.org/10.1017/jog.2020.34

Soheb, M., Ramanathan, A., Bhardwaj, A., & Sam, L. (2022). Spatiotemporal quantification of key environmental changes in Stok and Kang Yatze regions of Ladakh Himalaya, India. *Geocarto International*, 1–25. https://doi.org/10.1080/10106049.2022.2060312

Tacoli, C. (2017). *Migration and inclusive urbanization.* Department of Economics and Social Affairs, United Nations Secretariat. https://www.un.org/en/development/desa/population/events/pdf/expert/27/papers/V/paper-Tacoli-final.pdf

Thayyen, R. J., Dimri, A. P., Kumar, P., & Agnihotri, G. (2013). Study of cloudburst and flash floods around Leh, India, during August 4–6, 2010. *Natural Hazards, 65*(3), 2175–2204. https://doi.org/10.1007/s11069-012-0464-2

Tiwari, P. C. (2000). Land-use changes in Himalaya and their impact on the plains ecosystem: Need for sustainable land use. *Land Use Policy, 17*(2), 101–111. https://doi.org/10.1016/S0264-8377(00)00002-8

Tiwari, P. C., Tiwari, A., & Joshi, B. (2018). Urban growth in Himalaya: Understanding the process and options for sustainable development. *Journal of Urban and Regional Studies on Contemporary India, 4*(2), 15–27. https://doi.org/10.15027/45582

United Nations. (2015). *World population prospects: The 2015 revision, key findings and advance tables*. Department of Economic and Social Affairs, Population Division. https://population.un.org/wpp/publications/files/key_findings_wpp_2015.pdf

United Nations. (2019). *World urbanization prospects: The 2018 revision*. Department of Economic and Social Affairs, Population Division. https://population.un.org/wup/publications/Files/WUP2018-Report.pdf

United Nations. (2022). *World population prospects 2022: Summary of results*. Department of Economic and Social Affairs, Population Division. https://www.un.org/development/desa/pd/sites/www.un.org.development.desa.pd/files/wpp2022_summary_of_results.pdf

Verma, S., Sharma, A., Yadava, P. K., Gupta, P., Singh, J., & Payra, S. (2022). Rapid flash flood calamity in Chamoli, Uttarakhand region during Feb 2021: An analysis based on satellite data. *Natural Hazards, 112*(2), 1379–1393. https://doi.org/10.1007/s11069-022-05232-y

Wang, Y., Wu, N., Kunze, C., Long, R., & Perlik, M. (2019). Drivers of change to mountain sustainability in the Hindu Kush Himalaya. In P. Wester, A. Mishra, A. Mukherji, & A. B. Shrestha (Eds.), *The Hindu Kush Himalaya Assessment* (pp. 17–56). Springer International Publishing. https://doi.org/10.1007/978-3-319-92288-1_2

Wester, P., Mishra, A., Mukherji, A., & Shrestha, A. B. (Eds.). (2019). *The Hindu Kush Himalaya assessment: Mountains, climate change, sustainability and people*. Springer International Publishing. https://doi.org/10.1007/978-3-319-92288-1

World Bank. (2013). *Urbanization Beyond Municipal Boundaries: Nurturing Metropolitan Economies and Connecting Peri-Urban Areas in India*. The World Bank. https://doi.org/10.1596/978-0-8213-9840-1

Ziegler, A. D., Cantarero, S. I., Wasson, R. J., Srivastava, P., Spalzin, S., Chow, W. T. L., & Gillen, J. (2016). A clear and present danger: Ladakh's increasing vulnerability to flash floods and debris flows: Tourism and vulnerability to floods. *Hydrological Proccesses, 30*, 4214–4223. https://doi.org/10.1002/hyp.10919

Chapter 2
Regional Context of the Study Area

Abstract This chapter offers a comprehensive analysis of the geographical setting of the study area within the Ladakh region. It explores the region's unique topography, characterised by high-altitude terrain, rugged mountain ranges, and arid landscapes. The chapter also addresses the harsh climatic conditions, including extreme temperature variations, low precipitation, and the challenges posed by its cold desert ecosystem. Additionally, it examines the geographic and agricultural dynamics, highlighting the interplay between the challenging terrain and climate, as well as the adaptation strategies employed by local communities for farming practices and irrigation systems. The region's rich cultural heritage and diverse linguistic landscape are also highlighted, shaped by a long history of interactions between Tibetan Buddhist and Islamic traditions. Furthermore, it examines key economic activities that play a crucial role in sustaining the local economy. By integrating these aspects, the chapter offers a holistic understanding of Ladakh's geographical, cultural, and economic dimensions.

Keywords Climate · Culture · Economy · Agriculture · Muslim · Buddhist · Irrigation system · Kargil · Padum · Sankoo · Drass · Zanskar · Suru Valley · Ladakh

A. Hussain, *Urbanisation and Development Challenges in Kargil, Ladakh*, Advances in Asian Human-Environmental Research, https://doi.org/10.1007/978-3-032-05667-2_2

2.1 Geographical Setting

2.1.1 Ladakh

Ladakh[1] is a cold desert region located in the Indian Trans-Himalaya. It forms the westernmost part of the Tibetan Plateau, positioned in the northern section of the Western Himalayas. Geographically, the region is bordered by Tibet to the east, Himachal Pradesh to the south, the UT of Jammu and Kashmir to the west, and Pakistan-administered Gilgit-Baltistan to the north-west. Administratively, Ladakh comprises two districts: Leh, covering the central and eastern areas with a land area of 45,110 km^2, and Kargil, situated in the north-west with an area of 14,086 km^2. The Ladakh region of Leh district has most of the Buddhist population, while Kargil district has a majority of Shia Muslim population. The region is traversed by four prominent mountain ranges: the Great Himalaya, Zanskar, Ladakh, and Karakoram (Sapru, 2001). The great mountain valleys like Zanskar, Indus, Shyok, Suru, Drass, and Nubra are located among these major mountain ranges. The terrain is predominantly mountainous with sparse vegetation, and elevations vary considerably, ranging from 1000 to 7000 m above sea level (a.s.l.) (Fig. 2.1). Pangong Tso, Tsomoriri, and Tsokar are important lakes, and the rivers like Indus, Shyok, Hanle, Zanskar, Suru, and Drass are the major main sources of irrigation in Ladakh. The glaciers, snow cover, and permafrost are the main sources of surface and groundwater, which are harnessed through irrigation system for agriculture and human consumption. The agricultural productivity is low based on subsistence agriculture with restricted growing season between April and September. Besides subsistence agriculture, livestock such as sheep, goats, yaks, dzos, horses, ponies, and donkeys are domesticated for daily needs and domestic consumption.

The region is connected to the lowland by roads only during summer, while during winter only air connectivity is available because of heavy snowfall in Zoji La (pass) and Rohtang pass. In July 1979, the reorganisation of Ladakh district led to its division into two independent districts—Leh and Kargil—resulting in the formal creation of Kargil district (Sapru, 2001). The study area primarily focuses on Kargil town, which serves as the administrative centre of the region. However, the research also encompasses key urbanised settlements that are crucial for understanding the dynamics of the area. These include Padum in Zanskar Valley, Sankoo in Suru Valley, and Drass in Drass Valley. Kargil (2,676 m a.s.l.) and Sankoo (3,000 m a.s.l.) have lower elevations compared to Padum (3,670 m a.s.l.) and Drass (3,300 m a.s.l.), making them more favourable for cultivation and agriculture.

---

[1] Ladakh is the most commonly used name for the region today; however, it was historically known by other names such as Maryul (meaning Lowland), Marpoyul (Redland), and Khachanpa (Snowland) (Cunningham, 1854). These Tibetan terms were frequently used by early Chinese travellers. Presently, Ladakh is widely known as 'Little Tibet' or 'Western Tibet'. Interestingly, the term 'Little Tibet' originally referred to Baltistan, while Ladakh was called 'Great Tibet' by the Kashmiri Sultans who invaded the region in the fifteenth century (Petech, 1977).

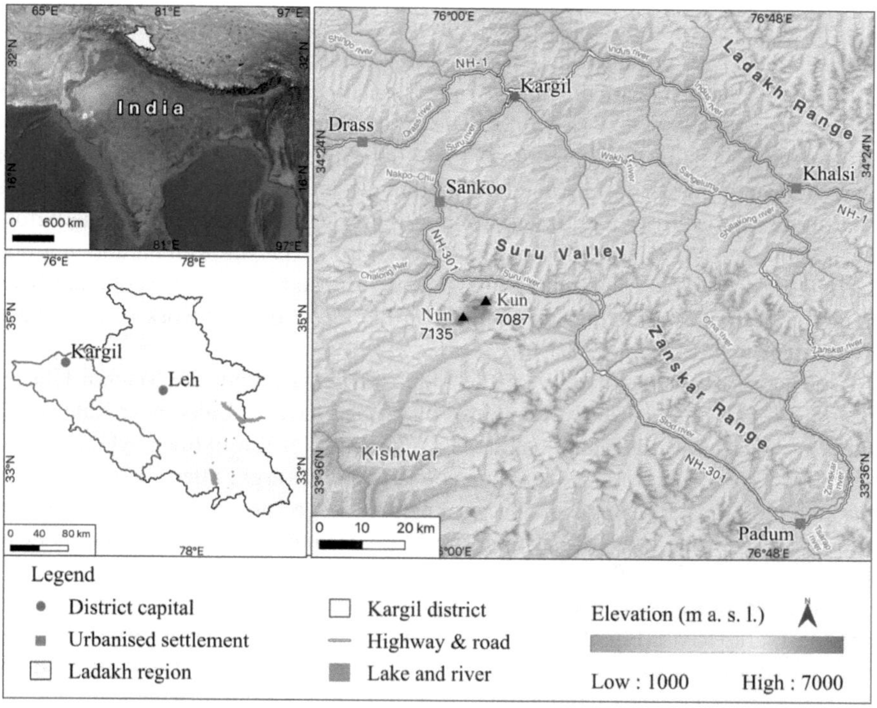

Fig. 2.1 Location of the study area in Kargil district of Ladakh, India. *Sources* SRTM (11 February 2000), OpenStreetMap (OSM). *Draft and cartography* Altaf Hussain

2.1.2 *Kargil District*

Kargil district[2] is located in the eastern part of the Kashmir region within the UT of Ladakh. It shares borders with Pakistan-administered Gilgit-Baltistan to the west, the UT of Jammu and Kashmir to the south-west, Himachal Pradesh to the south, and Leh district of Ladakh to the east. Geographically, it is separated from the Indian subcontinent by the Greater Himalayan Range, bounded by the Karakoram Range to the north, and largely encompasses the Zanskar Range. Shia Muslims constitute the majority of the local population, followed by Buddhists, Sunni Muslims, and Sikhs. The army population is mostly located in the border areas along the Line of Control (LOC) in Kargil district. The district is home to some of the highest mountain passes, including Zoji La (3,528 m a.s.l.), Fotu La (4,108 m a.s.l.), Namika La (3,700 m a.s.l.), and Penzi La[3] (4,400 m a.s.l.), which are among the highest in the region.

[2] Administratively, Kargil district comprises one municipal committee, 7 tehsils, 15 blocks, and 130 villages.

[3] Drang-Drung and Parkachik glaciers are located near Penzi La on the road (NH-301) between Kargil and Zanskar. Drang-Drung is perhaps the largest glacier after the Siachen glacier in Ladakh. The maximum length of the Drang-Drung glacier is 23 km and is the main source of the Stod River,

The district is composed of four high-altitude natural valleys: the Suru Valley, Indus Valley, Drass Valley, and Kanji Valley (India Meteorological Department, 2014). The main rivers are the Suru, Drass, Zanskar, and Wakha, along whose banks the majority of the population resides. The Suru River is formed from the confluence of several streams in the Rangdum plain (4,000 m a.s.l.). The river flows west for approximately 52 km from Rangdum to Tangole in the Suru Valley. It then turns sharply north towards Kargil town, where it joins the Drass River at Kharul, before eventually merging with the Indus River near Marol. The Nun Kun mountain group lies a few kilometers in direct line south-east of Suru, and the massif is comparatively small, and compact standing up by itself in the midst of a network of mountains (Workman & Workman, 1909).

The Drass River, a tributary of the Indus, originates from the Machoi Glacier near the Zoji La Pass. The river is joined by Shingo River at Latoo near Kaksar and flows northeast towards Kharul, situated 4.5 km north of Kargil town, where it joins the Suru River. All rivers and streams converge near Kargil, after which they pass through the Trans-Himalayan range via a narrow gorge known as the 'Wolf's Leap' and merge with the Indus River opposite Marol (Cunningham, 1854; Gazetteer of Kashmir and Ladák, 1890).

The Zanskar River is formed by the confluence of two tributaries—the Stod River and the Tsarap River (Fig. 2.1). The Stod (also known as Doda River) originates near the Pensi La at an altitude of 4,400 m a.s.l. and flows south-east towards Padum. Tsarap is formed by the merging of the tributaries of the Kurgiak River (also known as Kargyag Chu), originating near Shinku La and Tsarap River itself (also known as Lungnak River), originating near the Baralacha La. These tributaries converge below Purne village, forming the Tsarap River. The Tsarap River then flows north-west through a narrow gorge towards Padum, where it merges with the Stod River downstream of the town to form the main Zanskar River. The Zanskar River then proceeds to flow northward, eventually joining the mighty Indus River at Nimmu. This confluence is a crucial part of the Zanskar Valley's landscape and plays an essential role in the region's hydrology and ecology. Similarly, the Wakha River originates from the Rangdum plain in the Zanskar Range and flows north-west, passing through several villages, including Wakha, Mulbekh, Shargole, Lotsum, and Pashkyum. It eventually joins the Suru River close to Kargil town and becomes part of the Indus River system.

In Kargil, vegetation is sparse, largely treeless, and limited in herb and subshrub diversity. Dense greenery—such as Cyperaceae turf, meadows, cultivated crops (wheat, barley, potatoes), and deciduous tree plantations (poplars, willows, apricots)—is primarily confined to irrigated zones and streamside areas (Klimeš & Dickoré, 2005). The main crops of Kargil district consist of wheat (kro), barley (nas), pulses, and vegetables. The major common fruits of the area include apricots, apples, cherries, peaches, and grapes, and the best apricot fruit (halman) is grown in Karkitchu, Hardas, Shilikchey, and Batalik. Snow and glaciers are the main source

---

a major tributary of the Zanskar River. These glaciers' meltwater adds to the flow of the Suru and Stod rivers and plays a vital role in irrigating agricultural land.

of crop cultivation and agriculture. The agricultural fields are irrigated though local irrigation system by rivers, streams, canals, and other sources (Hill, 2014).

2.1.3 Kargil

Kargil, a statutory town, is situated on a fluvial terrace along the orographic left bank of the Suru River. It is the second-largest town in the UT of Ladakh after Leh and serves as the administrative and economic centre of Kargil district. Kargil municipality was declared as a Notified Area Committee (NAC) in 1963, and Municipal Committee (MC) of Kargil came into existence in 1979 from erstwhile NAC. Kargil town covers a municipal area of nearly 8.93 km^2 and comprises 13 municipal wards, stretching from Kako Shilikchey in the north to Titichumik in the south, and from Bagh-e-Khomeini in the east to Baroo Zgang in the west (Fig. 2.2). A series of flat terraces composed of alluvial gravel spans approximately 20 km^2 between the Suru and Wakha rivers. These terraces form important sites for urban expansion and development (Hussain et al., 2023a). The rugged terrain surrounding Kargil town ranges in elevation from approximately 2,600–4,500 m a.s.l., with steep slopes exceeding 75°. The area receives less than 300 mm of annual precipitation, most of which falls during the winter months, predominantly as snow (kah).

The district is prone to multiple natural disasters like cloudbursts and flash floods and witnessed such disasters recently in Baroo and Titichumik of Kargil town as well as in few villages of Suru, Drass, and Chiktan in 2022. Owing to the semi-arid conditions, agricultural production relies entirely on irrigation sourced from the Suru and Wakha rivers, as well as smaller tributaries fed by meltwater from glaciers and snow (Hill, 2014). The groundwater aquifer plays an important role as a water resource besides the surface water. The total population of the municipal area of Kargil was 10,657 in 2001, which increased to 18,558 in 2020 (Census of India, 2001; Municipal Committee Kargil, 2021). Total households registered in the municipal limit increased from 2,065 to 2,446 between 2001 and 2020 (Census of India, 2001; Municipal Committee Kargil, 2021). NH-1 originates from Kashmir, connects with NH-301 in Kargil town, and continues towards Leh, serving as the primary route linking the Kashmir Valley with the Ladakh region. Meanwhile, NH-301 begins in Kargil town, heading towards Zanskar, providing vital access to this remote region and facilitating transportation for both locals and tourists. Additionally, both highways are essential for the movement of goods, services, and emergency support throughout the region (Fig. 2.3).

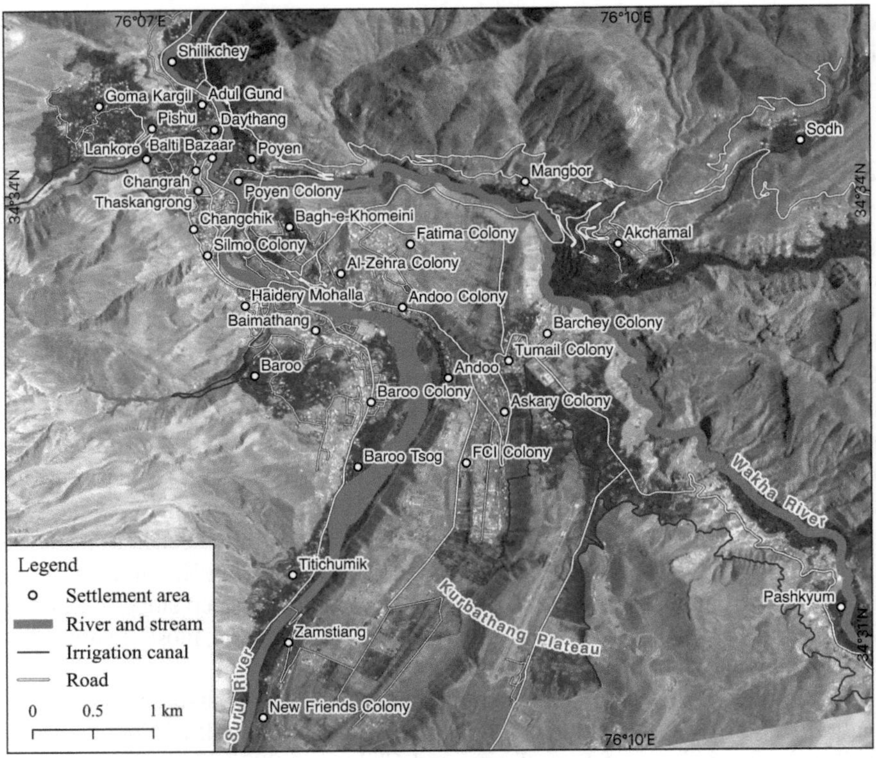

Fig. 2.2 Detailed distribution of settlement areas in Kargil town, Ladakh. *Sources* PlanetScope (8 June 2020); OpenStreetMap (OSM)

2.1.4 Padum

Padum, located at an altitude of 3,670 m a.s.l., is a small urbanised settlement of Zanskar region, south-east to Kargil. This area is known for its unique geographical features, with permanent habitation found between altitudes of 3,500 and 4,200 m a.s.l. Owing to the steep topography, human settlements are largely restricted to alluvial fans, elevated terraces, and the vast river plain located south of the meeting point of the Stod and Tsarap rivers (Deboos, 2012). Agriculture is still practised, covering an area of about 342 ha in 2011 (District Statistical & Evaluation Office Kargil, 2020). It largely depends on glacier and snow-fed irrigation systems combined with livestock farming. Cultivation remains productive but is constrained by a brief four-month growing season and limited irrigation resources (Dame & Mankelow, 2010). Padum changed in the past decade from being an agriculturally rural-oriented village (yul) to semi-urban characteristics (Crook et al., 1994). Moreover, there are 25 inhabited villages in Zanskar tehsil (District Statistical & Evaluation Office Kargil, 2021). The inhabitants are mainly Buddhists and Sunni Muslims with a population

Fig. 2.3 Urban structure of Kargil, Ladakh. The town is situated at an average altitude of 2,676 (m a.s.l.) on the bank of the Suru River. Kargil has developed into an administrative and commercial hub, as well as a centre for all development activities. Photograph taken by Altaf Hussain, 12 August 2021

of 13,793 in Zanskar tehsil as per the Census of India 2011. NH-301 connects Kargil with Padum, crossing challenging landscapes and high-altitude passes such as Pensi La, acting as a crucial link to the remote Zanskar region (Fig. 2.4).

2.1.5 Sankoo

Sankoo, located at an elevation of 3,000 m a.s.l., is a small urbanised settlement of the Suru Valley, which is 40 km away from the administrative capital of Kargil. It is situated on the left bank of the Suru River, where it meets the Nakpo-Chu (black water); the Nakpo-Chu stream comes from the west and flows into the Suru River on its left bank. The total population of Sankoo tehsil is 40,548 in Kargil district as per the Census of India 2011. The Suru Valley, including Sankoo, practices agriculture as well as the domestication of livestock in all households. The main agricultural produce consists of wheat (kro), barley (nas), peas, potatoes, and a small range of vegetables, while large pasturelands sustain livestock such as cattle, sheep, goats, and horses (Grist, 2008). The cultivation mostly relies on water irrigation system. The Suru River, Nakpo-Chu, and small streams play a significant role in irrigating agricultural fields. NH-301 passes through the central part of Sankoo in the Suru Valley, providing crucial connectivity to this picturesque, urbanised settlement (Fig. 2.5).

Fig. 2.4 Urbanised structure of Padum. The settlement is located at the junction of the Stod and Tsarap rivers, at an average altitude of 3,670 (m a.s.l.) in the Zanskar region. Photograph taken by Altaf Hussain, 26 August 2022

Fig. 2.5 Urbanised structure of Sankoo. The settlement lies at an average altitude of 3,000 (m a.s.l.) in the Suru Valley. The Suru River flows through the middle of Sankoo. Photograph taken by Altaf Hussain, 10 September 2022

2.1.6 Drass

Drass (Tibetan word meaning snow-fed) is a small urbanised settlement at an altitude of 3,300 m a.s.l. in the Drass Valley. It is the most westerly portion of the Kargil (Purig) (Gazetteer of Kashmir and Ladák, 1890). The population of Drass includes Baltis, Dards, a small number of Ladakhis, and a sect known as the Brokpa, with each group contributing to the region's unique cultural and ethnic diversity (Adair & Drew, 1875; Godfrey, 1899). In addition, there are 19 inhabited villages in Drass tehsil as per the census data (District Statistical & Evaluation Office Kargil, 2021). The combination of livestock farming and subsistence agriculture supported by canal irrigation systems forms the core element of the livelihood economy. Livestock farming includes cattle like ponies, cows, yaks, sheep, and goats. Barley (nas), wheat (kro), and buckwheat are the major cultivated crops combined with vegetable cultivation. NH-1 passes through Drass, serving as a vital link between Srinagar and Leh and playing a crucial role in regional connectivity (Fig. 2.6).

Fig. 2.6 Urbanised structure of Drass. The settlement is situated at an average elevation of 3,300 (m a.s.l.) in the Drass Valley, on the bank of the Drass River. Photograph taken by Altaf Hussain, 13 September 2022

2.2 Climate

Kargil district experiences a diverse climate ranging from sub-tropical in the lower areas to temperate and alpine conditions in the higher altitudes. The warmest period occurs in July and August, when average high temperatures reach around 28 °C and lows drop to approximately 13 °C. January is typically the coldest month, with average maximum temperatures near 2 °C and minimum temperatures falling to about −12 °C (India Meteorological Department, 2014). Situated in the Himalayan rain shadow, the district receives little rainfall as the monsoon winds lose most of their moisture while crossing the plains and the Himalayan ranges (India Meteorological Department, 2014). This climatic pattern is largely due to a significant depression in the Himalayas near the source of the Drass River, which enables the persistent humid air masses from Kashmir to pass northward. These vapours condense in the cold environment and result in either rain or snowfall, depending on the season (Gazetteer of Kashmir and Ladák, 1890). In summer, temperatures in the lower Indus Valley can exceed 30 °C, while in winter, Drass experiences extreme cold, with temperatures dropping below –40 °C (Passang et al., 2022). Snow accumulation reaches 71% snow-covered area in the entire region of Ladakh between September and February (Passang et al., 2022).

During the time period 1981–2022, the highest monthly mean precipitation was 36.1 mm in March, followed by April, February, and January, while the lowest monthly mean precipitation was 14.2 mm in October, which is followed by September, Decemeber, and Novermber (Fig. 2.7). Kargil district receives an average annual precipitation of 319.4 mm (India Meteorological Department, 2014). Of this, only 24% occurs during the monsoon season (June to September), while 33% falls in the pre-monsoon period (March to May). March records the highest average rainfall (charpha) at 44.2 mm. The winter season (December to February) accounts for approximately 31% of the total annual precipitation (India Meteorological Department, 2014).

Between 1981 and 2022, the highest total annual precipitation of 532.6 mm was measured in 2015, which was followed by 522.1 mm in 2019, 475.8 mm in 2022, and 438.2 mm in 2021 (Fig. 2.8), while the lowest total annual precipitation of 84.4 mm was registered in 2000, 89.7 mm in 1999, and 100.2 mm in 1991. The plot shows the fluctuation in the total annual precipitation with sharp decrease from 1997 to 2000, and then an increasing trend since 2001. The highest annual mean maximum temperature was 25.3 °C in 2001 and the lowest annual mean maximum temperature was 21.4 °C in 1987 and 2016. On the other hand, the coldest annual mean minimum temperature of –38.4 °C was recorded in 2008, and lowest annual mean minimum temperature was –22 °C in 1998.

The Ladakh region has recently experienced natural hazards due to extreme precipitation events. In Kargil, recurring flash floods were observed in 2006, 2010, 2013, 2014, 2016, 2018, and 2022, causing massive damage to roads, buildings, and agricultural areas (Hussain et al., 2023b). In 2022, the flash floods affected Kargil town (Titichumik and Baroo) as well as rural villages in Suru, Chiktan, and Drass. In 2010, the capital town of Leh was hit by severe cloudburst floods, which caused 255

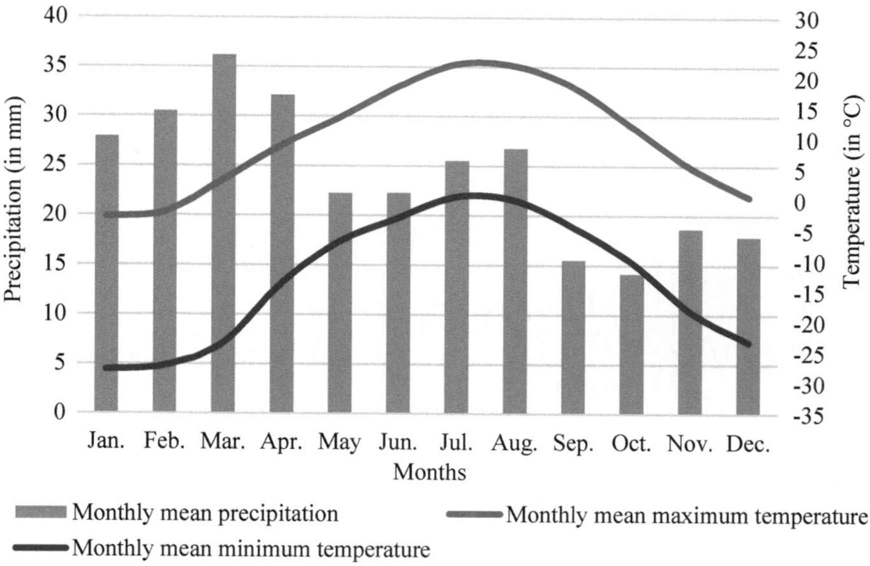

Fig. 2.7 Monthly mean precipitation and monthly mean maximum/minimum temperature in Kargil from 1981 to 2022. *Sources* NASA Langley Research Centre POWER Project funded through the NASA Earth Science Directorate Applied Science Program (https://power.larc.nasa.gov/data-access-viewer/)

deaths of people and the destruction of roads, bridges, hospitals, and human settlements (Thayyen et al., 2013). Climate change has triggered the occurrence of glacial lake outburst floods (GLOFs), posing significant threats to communities and settlements situated along glacial streams in Ladakh (Schmidt & Nüsser, 2023; Schmidt et al., 2020). Shifts in precipitation patterns and climate warming may have significant impacts on the region's ecology, vegetation, hydrology, cryosphere, wildlife, and human society (Chevuturi et al., 2018).

2.3 Culture

Ladakh is culturally diverse and has nurtured a rich and evolving heritage over time. This cultural legacy is reflected in its monuments, artistic traditions, monasteries (gompas), oral literature, festivals, and the communal celebration of various events (Sapru, 2001). Significant life and agricultural events—such as marriages, births, sowing and harvesting, and the blooming of plants—are celebrated with enthusiasm, featuring feasts, dances, and folk songs that highlight the region's vibrant culture (Sapru, 2001). The Ladakh region is home to a diverse mix of ethnic groups, such as the Purigpa, Tibetans (Ladakhi), Baltis, Mons, Dards, and Changpas. The present-day Ladakhis are considered to be descendants of a diverse mix of ethnic groups, including the Mons from northern India, the Dards of Baltistan, and Central Asian Mongols. Historical accounts suggest that the Mons were the earliest settlers,

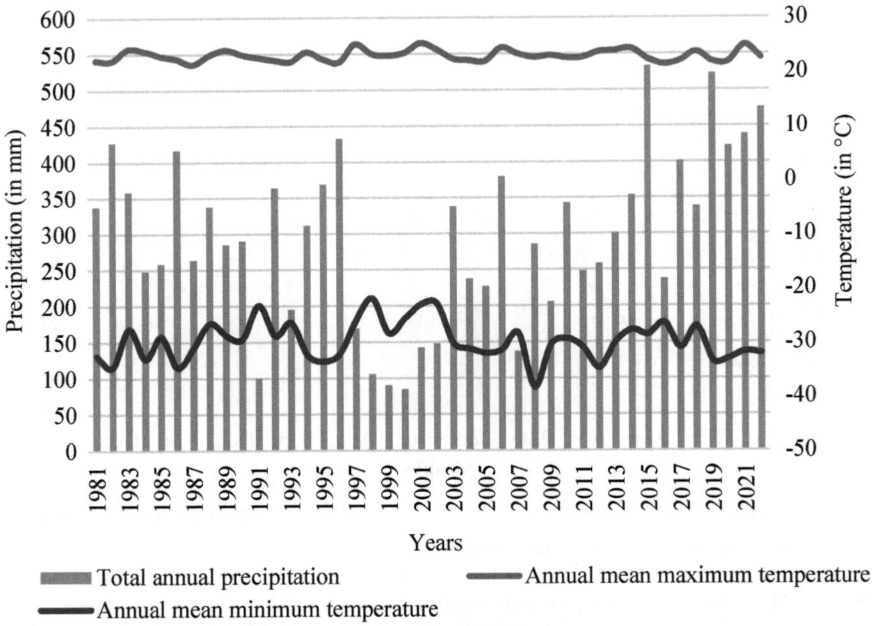

Fig. 2.8 Total annual precipitation and annual mean maximum/minimum temperature from 1981 to 2022. *Sources* NASA Langley Research Centre POWER Project funded through the NASA Earth Science Directorate Applied Science Program (https://power.larc.nasa.gov/data-access-viewer/)

succeeded by the Dards who established settlements along the Indus Valley. The Mongol migration to Ladakh is believed to have occurred in the early tenth century A.D. (Zutshi et al., 1961). The Mon community is traditionally known for its role as professional entertainers, travelling from place to place and performing with musical instruments. The Dards[4] are predominantly settled in the Drass and along the Indus Valley. The Tibetans (Ladakhis) are predominant in Eastern and Central Ladakh and have gradually merged with other ethnic groups over time to form a homogenous Ladakhi identity (Sapru, 2001). Among the ethnically and culturally distinct groups in Ladakh are the Tibetans proper, who reside in Choglamsar, and the nomadic Changpas of Changthang, known for herding Pashmina goats. The Balti population, including the Purigpas, is predominantly found in Western Ladakh (Kargil), with smaller, scattered communities also present in the Nubra Valley and areas around Leh (Sapru, 2001). The region is inhabited by Muslims (46%), Buddhists (40%), Hindus (12%), Sikhs (0.8%), and Christians (0.5%). The total population of Ladakh

[4] The Brokpas (Drokpas) are the remaining descendants of Ladakh's Dard immigrants. The Brokpas of the Drass became Muslims after the mid-fifteenth century, while the Brokpas of the lower Indus villages have neither accepted Islam nor assimilated into the mainstream of Ladakh's population (Schettler & Schettler, 1989). The inhabitants follow a faith that is a mixture of Animism and Shamanism. The most popular festival 'Bonona' is still celebrated every three years at Garkon and Da villages, and it involves the ritual slaughtering of goats at the altar of their deities.

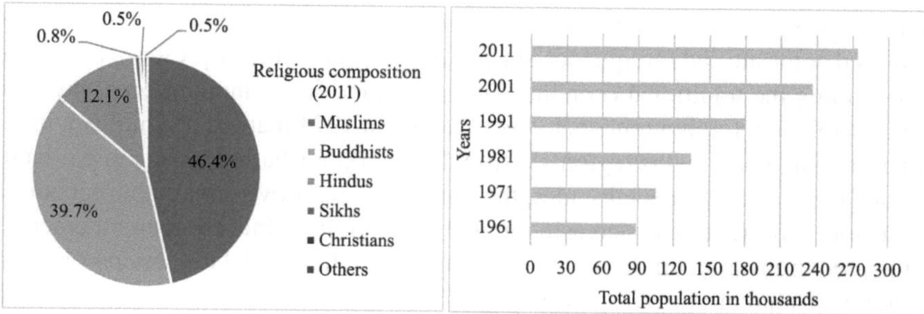

Fig. 2.9 Total population and religious composition of Ladakh, India. *Sources* Census of India (1961, 1971, 1981, 2001, and 2011)

was 88,651 in 1961, which considerably increased to 179,410 in 1991 and 274,289 in 2011, according to the Census of India (Fig. 2.9).

The predominant religion in Ladakh is Vajrayana Buddhism, a sect of Mahayana Buddhism, which incorporates elements of the ancient Bon animistic tradition and Hindu Tantric practices. Buddhism is the predominant religion in Central and Eastern Ladakh, whereas Shia Islam is primarily followed in Western Ladakh (Sapru, 2001). Ladakh celebrates several socio-religious festivals throughout the year, with Eid-ul-Fitr and Losar being the most significant. During these festivals and ceremonies, people exchange greetings, gifts, and ceremonial scarves (khataks). The language spoken throughout Ladakh is essentially Tibetan, but there is a great variety of dialects in the language. Bhoti language is mostly spoken in Leh, while Purigi language is mostly spoken by people of Kargil. Tibetan language of Balti is predominantly spoken in the Gilgit-Baltistan. It is also spoken in the areas of Nubra Valley (Turtuk, Bogdang, Tyakshi), and in the border areas of Kargil (Hardas, Karkitchu, Lato). The Dardic language of Shina is widely spoken in the Drass Valley. In the Ladakh region, particularly in Dah, Hanu, Garkon, and Darchik, the Brokskat language is spoken, which is considered a divergent variety of the Shina language. The Zanskari dialect of Tibetan language is mostly spoken in Zanskar region. Bodhi, Bhoti, Tibetan, etc. were the different names of languages used for what has been categorically termed as Ladakhi (Khan, 1912).

2.4 Economy

Ladakh's economy is primarily based on agriculture, livestock rearing, tourism, military presence, and government employment opportunities. Traditionally, Ladakh region has relied on subsistence agriculture, livestock farming, and crop cultivation such as barley (nas), wheat (kro), peas, and vegetables. Animal husbandry, which encompasses the rearing of livestock like yaks, sheep, goats, and dzos, is also a crucial component of the economy and provides essential resources like milk, meat, wool, and transportation (Dame, 2018; Dame & Nüsser, 2011; Dollfus, 2013; Ladon et al., 2023). There are ongoing changes in the economy of Ladakh. The advancement of irrigation systems, innovation in farming techniques, improved road connectivity, increased domestic and foreign tourism, and a significant military presence have been transforming the traditional economy into an income-based modern economy. The decision to allow foreign tourists to visit Ladakh in 1974 marked a significant milestone, shaping not only the region's history but also contributing to the rapid urbanisation. Over the subsequent decades, the tourism sector drove remarkable economic expansion, leveraging the increasing global tourist influx. Indeed, tourism emerges as the predominant factor driving the continual growth of Ladakh's service-oriented economy (Dame et al., 2019; Müller, 2022). It has gained significant recognition as a crucial industry in Ladakh due to its potential for creating employment opportunities and generating substantial income. A significant portion of the population is employed in allied sectors such as transportation, hospitality, food services, and small-scale industries (Dame et al., 2019; Hussain et al., 2023a; Müller et al., 2020; Vogel & Field, 2020). This industry directly impacts the socio-economic landscape of the region. Additionally, the industry promotes cultural exchange and helps in the preservation of local traditions and heritage.

The Ladakh region's stunning landscapes, cultural heritage, and adventure opportunities attract tourists from across India and around the world contributing significantly to the local economy of Ladakh. Popular tourist destinations such as Suru Valley, Nubra Valley, Markha Valley, Indus Valley, Drass Valley, and Zanskar Valley offer a range of activities including trekking, mountaineering, and river rafting. Additionally, Ladakh is strategically important due to its proximity to the borders with China and Pakistan, resulting in a significant military presence in the region (Baghel & Nüsser, 2015; Gupta, 2022). This military presence not only ensures security but also influences local economic dynamics through infrastructure development and employment opportunities (Baghel & Nüsser, 2015). For centuries, the Ladakh thrived on subsistence agriculture, pastoralism, and caravan trade, fostering a self-reliant way of life. However, in recent decades, there has been a discernible shift towards adopting external influences. Government initiatives advocate for modernisation and globalisation to enhance the economy. As a result, there is an increasing reliance on the broader Indian and international economies (Dame, 2018; Dame & Nüsser, 2008, 2011). The recent creation of the UT of Ladakh in 2019 has brought new opportunities for the region's economy such as renewable energy, promotion of local industries, fostering entrepreneurship, skill development, and infrastructural development. However, it is crucial to ensuring that development initiatives are carried out in a manner that respects Ladakh's fragile ecosystem and cultural heritage.

References

Adair, F. E. S., & Godfrey, S. H. (1899). *A summer in high Asia: Being a record of sport and travel in Baltistan and Ladakh, with an appendix on central Asian trade, illustrated from drawings by the author, photographs, and a map of the route*. W. Thacker & Company.

Baghel, R., & Nüsser, M. (2015). Securing the heights: The vertical dimension of the Siachen conflict between India and Pakistan in the Eastern Karakoram. *Political Geography, 48*, 24–36. https://doi.org/10.1016/j.polgeo.2015.05.001

Census of India. (1961). *Primary census abstract*. Office of the Registrar General & Census Commissioner, India. https://censusindia.gov.in/

Census of India. (1971). *Primary census abstract*. Office of the Registrar General & Census Commissioner, India. https://censusindia.gov.in/

Census of India. (1981). *Primary census abstract*. Office of the Registrar General & Census Commissioner, India. https://censusindia.gov.in/

Census of India. (2001). *Primary census abstract*. Office of the Registrar General & Census Commissioner, India. https://censusindia.gov.in/

Census of India. (2011). *Primary census abstract*. Office of the Registrar General & Census Commissioner, India. https://censusindia.gov.in/

Chevuturi, A., Dimri, A. P., & Thayyen, R. J. (2018). Climate change over Leh (Ladakh), India. *Theoretical and Applied Climatology, 131*(1–2), 531–545. https://doi.org/10.1007/s00704-016-1989-1

Crook, J., & Osmaston, H. (Eds.). (1994). *Himalayan Buddhist villages: Environment, resources, society and religious life in Zangskar, Ladakh* (1st ed.). Motilal Banarsidass Publishers Private Limited.

Cunningham, A. (1854). *Ladák, physical, statistical, and historical: With notes of the surrounding countries*. Wm. H. Allen and Company.

Dame, J. (2018). Food security and translocal livelihoods in high mountains: Evidence from Ladakh, India. *Mountain Research and Development, 38*(4), 310. https://doi.org/10.1659/MRD-JOURNAL-D-18-00026.1

Dame, J., & Mankelow, J. S. (2010). Stongde revisited: Land-use change in Central Zangskar. *Erdkunde, 64*(4), 355–370. https://doi.org/10.3112/erdkunde.2010.04.05

Dame, J., & Nüsser, M. (2008). Development perspectives in Ladakh, India. *Geographische Rundschau International Edition, 4*(4), 20–27. https://www.researchgate.net/profile/Marcus-Nuesser/publication/279175071_Dame_und_Nuesser_2008_GRIntern/links/558bb5ea08ae02c9d1f968c3/Dame-und-Nuesser-2008-GRIntern.pdf

Dame, J., & Nüsser, M. (2011). Food security in high mountain regions: Agricultural production and the impact of food subsidies in Ladakh, Northern India. *Food Security, 3*(2), 179–194. https://doi.org/10.1007/s12571-011-0127-2

Dame, J., Schmidt, S., Müller, J., & Nüsser, M. (2019). Urbanisation and socio-ecological challenges in high mountain towns: Insights from Leh (Ladakh), India. *Landscape and Urban Planning, 189*, 189–199. https://doi.org/10.1016/j.landurbplan.2019.04.017

Deboos, S. (2012). Religious fundamentalism in Zanskar, Indian Himalaya. *Himalaya, the Journal of the Association for Nepal and Himalayan Studies, 32*(1), 35–42. http://digitalcommons.macalester.edu/himalaya/vol32/iss1/12

District Statistical & Evaluation Office Kargil. (2020). *Village amenity directory*. Directorate of Economics & Statistics, Planning Development & Monitoring Department.

District Statistical & Evaluation Office Kargil. (2021). *Statistical handbook*. Directorate of Economics & Statistics, Planning Development & Monitoring Department.

Dollfus, P. (2013). Transformation processes in Nomadic Pastoralism in Ladakh. *Himalaya, the Journal of the Association for Nepal and Himalayan Studies, 32*(1), 15. http://digitalcommons.macalester.edu/himalaya/vol32/iss1/15

Drew, F. (1875). *The Jummoo and Kashmir territories: A geographical account*. Edward Stanford.

Gazetteer of Kashmir and Ladák. (1890). *Together with routes in the territories of the Maharája of Jamú and Kashmír; compiled under the direction of the Quarter Master General in India in the Intelligence Branch*. Superintendent of Government Printing, India.

Grist, N. (2008). Urbanisation in Kargil and its effects in the Suru Valley. In M. van Beek, & F. Pirie (Eds.), *Modern Ladakh: Anthropological perspectives on continuity and change*. Brill. https://doi.org/10.1163/ej.9789004167131.i-313

Gupta, R. (2022). *Freedom in captivity: Negotiations of belonging along Kashmir's frontier*. Cambridge University Press.

Hill, J. (2014). Farmer-managed irrigation systems in Baltistan and Kargil. *Ladakh Studies, 31*, 4–23.

Hussain, A., Schmidt, S., & Nüsser, M. (2023a). Dynamics of mountain urbanisation: Evidence from the Trans-Himalayan town of Kargil, Ladakh, India. *Land, 12*(4), 920. https://doi.org/10.3390/land12040920

Hussain, A., Schmidt, S., & Nüsser, M. (2023b). *Urban landscape change in the Trans-Himalayan town of Kargil, Ladakh, India*. https://doi.org/10.5194/egusphere-egu23-2663

India Meteorological Department. (2014). *Climate of Jammu and Kashmir*. Additional Director General of Meteorology (Research), India Meteorological Department, Government of India. https://diragrijmu.nic.in/CSS%20GUIDELINES/FINAL_CLIMATE%20OF%20JAMMU%20AND%20KASHMIR_e%20book.pdf

Khan, M.-Z. (1912). *Census of India, 1911: Kashmir, Part I-Report, Vol. XX*. Newul Kishore Press.

Klimeš, L., & Dickoré, B. (2005). A contribution to the vascular plant flora of Lower Ladakh (Jammu & Kashmir, India). *Willdenowia, 35*(1), 125. https://doi.org/10.3372/wi.35.35110

Ladon, P., Nüsser, M., & Garkoti, S. C. (2023). Mountain agropastoralism: Traditional practices, institutions and pressures in the Indian Trans-Himalaya of Ladakh. *Pastoralism, 13*(1), 30. https://doi.org/10.1186/s13570-023-00289-1

Müller, J. (2022). *Urban mountain waterscapes in Leh, Indian Trans-Himalaya: The transformation of hydro-social relations*. Springer International Publishing. https://doi.org/10.1007/978-3-031-18249-5

Müller, J., Dame, J., & Nüsser, M. (2020). Urban mountain waterscapes: The transformation of hydro-social relations in the trans-Himalayan town Leh, Ladakh, India. *Water, 12*(6), 1698. https://doi.org/10.3390/w12061698

Municipal Committee Kargil. (2021). *Information document provided by Municipal Committee Kargil*. Ladakh Autonomous Hill Development Council, Kargil.

Passang, S., Schmidt, S., & Nüsser, M. (2022). Topographical Impact on Snow Cover Distribution in the Trans-Himalayan Region of Ladakh, India. *Geosciences, 12*(8), 311. https://doi.org/10.3390/geosciences12080311

Petech, L. (1977). *The Kingdom of Ladakh C. 950–1842 A.D.* Istituto Italiano per il Medio ed Estremo Oriente.

Sapru, C. S. (2001). *District census Handbook, Leh district: Village and town directory, series-2, part-A & B*. Deputy Director of Census Operations, Jammu & Kashmir.

Schettler, M., & Schettler, R. (1989). *Kashmir, Ladakh & Zanskar: A travel survival kit* (3rd ed.). Lonely Planet.

Schmidt, S., & Nüsser, M. (2023). Glaciers of central Ladakh: Distribution, changes and relevance in the Indian Trans-Himalaya. In B. Humbert-Droz, J. Dame, & T. Morup (Eds.), *Environmental change and development in Ladakh, Indian Trans-Himalaya* (pp. 11–30). Springer Nature Switzerland. https://doi.org/10.1007/978-3-031-42494-6_2

Schmidt, S., Nüsser, M., Baghel, R., & Dame, J. (2020). Cryosphere hazards in Ladakh: The 2014 Gya glacial lake outburst flood and its implications for risk assessment. *Natural Hazards, 104*(3), 2071–2095. https://doi.org/10.1007/s11069-020-04262-8

Thayyen, R. J., Dimri, A. P., Kumar, P., & Agnihotri, G. (2013). Study of cloudburst and flash floods around Leh, India, during August 4–6, 2010. *Natural Hazards, 65*(3), 2175–2204. https://doi.org/10.1007/s11069-012-0464-2

Vogel, B., & Field, J. (2020). (Re)constructing borders through the governance of tourism and trade in Ladakh, India. *Political Geography, 82*, 102226. https://doi.org/10.1016/j.polgeo.2020.102226

Workman, F. B., & Workman, W. H. (1909). *Peaks and glaciers of Nun Kun: A record of Pioneer exploration and mountaineering in the Punjab Himalaya*. Constable and Company Ltd.

Zutshi, J. N., Farooqi, B. A., Din, G. M., & Nanda, J. K. (1961). *Census of India: Village survey monograph of Kharboo (Tehsil Kargil, District Ladakh), Part VI, No. 13, Vol. VI*. Director of Census Operations, Jammu & Kashmir.

Chapter 3
Methodological Approach and Data Collection

Abstract This chapter outlines the methodological approach adopted for the selected study area in Ladakh. To comprehensively analyse and interpret the study area, a mixed-methods approach combined with multi-temporal and multi-scale remote sensing is employed. This approach integrates various sources of data such as statistical information, satellite imagery spanning different time periods and resolutions, field surveys, qualitative insights from interviews, direct observations, narratives gathered from local communities, photography, and other relevant datasets. Both quantitative and qualitative data are analysed using appropriate methodologies and tools to ensure the validity of statistical findings and to derive deeper insights through data interpretation. The supervised method and digitisation process are employed to investigate, quantify, and analyse land cover dynamics and detect changes across different time periods in the selected study area in Ladakh. This comprehensive approach seeks to provide deeper insights into the interactions between natural processes and human activities, contributing valuable knowledge for environmental management and sustainable development in the region.

Keywords Mixed-methods · Statistical data · Field surveys · Interviews · Photography · Municipal wards · Remote sensing · Satellite imagery · Supervised classification · Kargil · Ladakh

3.1 Mixed-Methods Research

Mixed-methods are useful for studying urbanisation and development challenges, as they enable researchers to develop a comprehensive understanding of complex urban dynamics. This approach offers a robust framework by combining both quantitative and qualitative techniques (Creswell, 1999; Creswell & Clark, 2017; Johnson et al., 2007; Silverman & Patterson, 2022; Teddlie & Tashakkori, 2009). The integration of quantitative data with qualitative narratives can offer a holistic understanding of urban dynamics, ensure data reliability, and help uncover the root causes

A. Hussain, *Urbanisation and Development Challenges in Kargil, Ladakh*, Advances in Asian Human-Environmental Research, https://doi.org/10.1007/978-3-032-05667-2_3

of urbanisation through various perspectives. Mixed-methods offer multiple perspectives and add layers of complexity to research that cannot be captured by a single method alone. They enable the understanding of complex phenomena qualitatively, while explaining the phenomena through numbers, charts, and statistical analyses (Creswell, 1999; Silverman & Patterson, 2022). Qualitative research focuses on exploring the meanings and lived experiences that shape human lives and their surrounding social contexts (Fossey et al., 2002).

It involves analysing data gathered through direct field observations, open-ended and in-depth interviews, and written records. The approach relies on naturalistic inquiry, exploring real-life contexts inductively to generate detailed narrative accounts and develop case studies (Patton, 2005). In contrast, quantitative research employs methods that systematically examine social phenomena through statistical data. It is grounded in the assumption that the subject of research study can be quantified and measured (Watson, 2015). A mixed-methods framework can lead to better inferences, and incorporate diverse views and ideas (Teddlie & Tashakkori, 2009). This approach involves the simultaneous application of two or more methods to investigate a research question at the same stage, within the same setting, and involving the same research participants (Philip, 1998).

According to Morgan, qualitative research generally focuses on goals that are inductive, subjective, and context-specific, whereas quantitative research tends to pursue objectives that are deductive, objective, and broadly generalisable (Morgan, 2013). The key distinction between these two approaches lies in their underlying logic: quantitative research is typically aligned with hypothesis formulation and testing through a deductive framework, whereas qualitative research follows an inductive approach focused on exploring patterns and meanings. This dichotomy is also reflected in the way research findings are structured and communicated in scientific versus exploratory academic writings (Philip, 1998). Thus, a methodology based on mixed methods is a highly suitable approach to collecting and obtaining empirical data and evidence for a research study (Fig. 3.1). The mixed-methods used and applied during this research study include interviews, discussions, field notes, observations, photography, district statistical data, census data, and document collection.

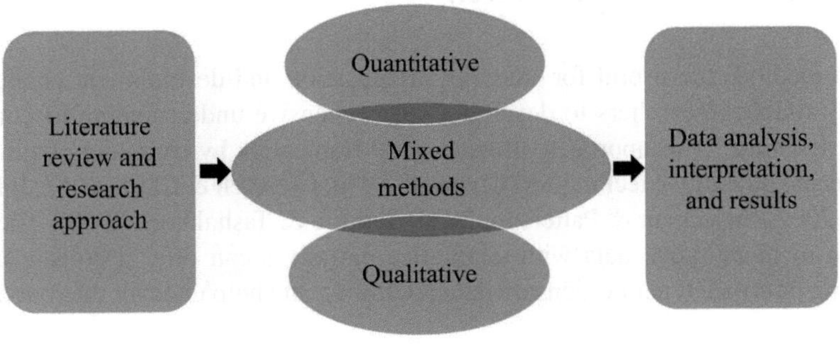

Fig. 3.1 Methodology of mixed-methods used for research investigation. Draft by Altaf Hussain

Data analysis is a nuanced and iterative process that requires navigating between concrete data and abstract concepts, integrating inductive and deductive reasoning, and balancing descriptive accounts with interpretive insights (Merriam & Tisdell, 2015). Qualitative analysis can be applied to a wide range of secondary data sources, such as print media content, institutional and organisational documents, website materials, and archival records (Silverman & Patterson, 2022). Additionally, it typically relies on observations of individuals' interactions within physical and social environments, as well as the examination of information derived from public records, historical documents, and other existing materials. The data collection of field study visits includes fieldwork, interviews, discussions, field notes, observations, and photography. Fieldwork was carried out during the summer and autumn months in 2021, 2022, 2023, and 2024 in the Kargil district. The included months were June, July, August, September, and October. The fieldwork was originally planned to start in 2020, but it was delayed due to COVID-19 pandemic and international travelling restrictions. Qualitative interviews were conducted to gather data, combined with ground truth field observations, to address the research questions. Interviews and face-to-face meetings were organised with LAHDC-Kargil officers, stakeholders of various institutions, and local people of Kargil in Ladakh.

Interviews were conducted using both unstructured and semi-structured formats. Unstructured interviews follow a natural, conversational style and are typically employed when the researcher lacks sufficient knowledge about a phenomenon to frame specific questions. With no predetermined set of questions, they serve primarily as exploratory tools (Bloomberg & Volpe, 2019; Merriam & Tisdell, 2015). In contrast, semi-structured interviews allow for a more focused examination of a particular topic. They consist of a blend of structured and open-ended questions that are flexibly worded, enabling the researcher to collect specific information while remaining adaptable during the conversation (Merriam & Tisdell, 2015). This format is especially useful for gaining in-depth insights into how various stakeholders—such as community residents, policymakers, social workers, and public administrators—perceive and understand issues within professional contexts (Silverman & Patterson, 2022).

Semi-structured interviews were used in this study to collect relevant data and information (Appendix A). This method was chosen because it provides flexibility, depth of understanding, and the ability to explore diverse perspectives, as well as contextual and qualitative insights about past and existing urban challenges and issues. These elements are essential for addressing the complexities of the urban environment. The interviews were conducted to understand and document the urban infrastructure challenges and urbanisation processes of Kargil town. Participants included residents of Kargil town, representatives from LAHDC-Kargil, community activists, officials, and professionals working in sectors related to urban development. The aim was to gather a comprehensive range of perspectives and insights from individuals involved in or affected by urban development initiatives and challenges. The participants included engineers, architects, women, teachers, students, hoteliers, businessmen, shopkeepers, vendors, and local people. Interviews were conducted in various settings, such as offices, homes, schools, and a few randomly in the main bazaar of Kargil town. All information was provided orally in Purgi, Balti, and Urdu. As a native of Kargil, I faced no language barriers during the fieldwork, which facilitated smooth communication. The questions were kept flexible and tailored according to the respondent's knowledge and expertise. All responses and narratives were documented for subsequent qualitative analysis. Field notes—comprising detailed descriptions of observations made in the field along with analytic reflections—serve as one of the most essential techniques in qualitative data collection. While field notes can be used independently, they are often employed in conjunction with other methods, such as semi-structured interviews and focus group discussions, to enrich the depth and context of the research (Silverman & Patterson, 2022). The interviews provided a detailed account and understanding of socio-economic issues, urbanisation processes, and the current state of urban development in Kargil town.

During the data collection process, brief meetings were conducted with executive officers, superintending engineers, executive engineers, and junior engineers from various departments, including the MC, Public Health Engineering (PHE), Pradhan Mantri Gram Sadak Yojana (PMGSY), Irrigation and Flood Control (I&FC), Roads and Buildings (R&B), Public Works Department (PWD), and Kargil Development Authority (KDA) under LAHDC-Kargil. Discussions primarily focused on urban infrastructure challenges and issues, as well as ongoing urban schemes. The MC office of Kargil was visited several times for gathering information and data. Formal discussions were held with the executive officer (Shafaqat Ali Khan) of MC and his staff members. A detailed personal discussion was conducted with Mohammad Abass, a junior engineer of MC Kargil, regarding the responsibilities and functions of MC in the urban development of Kargil. Similarly, data was also collected from the MC office of Kargil, including information on municipal wards, deep drains, solid waste management, public toilets, and urban infrastructure schemes.

Observation is one of the most common and important techniques of qualitative research (Bloomberg & Volpe, 2019; Silverman & Patterson, 2022). In this study, it was employed to assess the functioning of urban infrastructure and public services in Kargil town. In the municipal area of Kargil, urban infrastructure deficiencies were identified by observing the condition of water supply systems, toilets, deep drains, solid waste management, roads, traffic congestion, and public spaces. Observation enables researchers to identify and interpret complex interactions within natural social environments (Bloomberg & Volpe, 2019). Gaining such insight often requires extended engagement in the field, during which researchers may assume different roles to develop a deeper understanding of the individuals and phenomena under investigation (Baker, 2006).

3.2 Statistical Data

The quantitative data includes statistical data on demographic aspects, occupational distribution, transport, tourism, and climate.

Data on demographic characteristics and occupational distribution were collected, organised, and analysed using sources from the Office of the Registrar General and Census Commissioner, India, and the District Statistical and Evaluation Office, Kargil, UT of Ladakh (Census of India, 1961, 1971, 1981, 2001, 2011c; District Statistical & Evaluation Office Kargil, 2012, 2016, 2018, 2019, 2020a, 2020b, 2021, 2022, 2023, 2024). For demographic aspects, the census data from 1961 to 2011 were used to show the urban population, decade-wise population change, and level of urbanisation in Kargil town, combined with data from Kargil district and the Ladakh region. This comprehensive examination provided insights into urban population dynamics, the rate of growth over each decade, and the evolving level of urbanisation within these areas. By scrutinising these census data, the study sought to examine how urbanisation has unfolded in Kargil, the surrounding district, and its broader impact on the demographic landscape of Ladakh over the past six decades. The formula $P_n = P_o + N \times R$ is a simple linear projection used for estimating future population values based on an initial population (P_o), the number of periods (N), and a growth rate (R). Here, P_n represents the projected population at time n. This formula was specifically applied to estimate the 1991 population data[1] for Kargil town. Additionally, census data from 1971, 1981, 2001, and 2011 were utilised to calculate and compare the total population of Kargil, Zanskar, Sankoo, and Drass blocks. This analysis aimed to understand the demographic changes within these administrative divisions over the specified time frame. By utilising demographic information from the census records, the total population residing in each of these blocks was computed for the respective census years. This allowed for a detailed examination of population trends and fluctuations over the decades under study. Ultimately, the demographic figures for Kargil town and the blocks, as well as the entire Kargil district and Ladakh region, were included in the analysis. This broader scope provided insights into population dynamics at various administrative levels within the Ladakh region. The census data of 1981 and 2011 from the Census of India were used to analyse the population size of settlements within the Kargil district. This analysis provided insights into the settlement distribution of inhabitants across all villages in the district. It also helped in understanding the spatial distribution of population density and highlighted patterns of rural settlements. Furthermore, the Census of India 2011 was employed to examine the religious composition in Ladakh.

The census data from 1981 and 2011 were employed to analyse the occupational distribution of both main and marginal workers in Kargil town and the urbanised settlements of Padum, Sankoo, and Drass. To gain deeper insights into the occupational structure, the census data of 2001 and 2011 were utilised. Specifically, these data were used to analyse the distribution of main and marginal workers across different blocks within the Kargil district. This examination of the occupational distribution allowed for the identification of trends and patterns in the types of employment prevalent within the district over the course of a decade. According to the Census of

[1] Due to insurgency and political unrest, the 1991 Census was not carried out in Jammu and Kashmir. As a result, the 1991 population data for Kargil town were estimated by the author through linear projection methods.

India, main workers are classified into four groups: cultivators, agricultural labourers, household industry workers, and other workers (Census of India, 2011b). An individual who is engaged in any economically productive activity for six months or more during a year is classified as a main worker. In contrast, a person who works for more than three months but less than six months in a year is considered a marginal worker. An individual is considered a cultivator if they are involved in the cultivation of land owned by them or accessed from the government, private owners, or institutions in exchange for money, produce, or a share of the crop. Agricultural labourers are those who work on someone else's land for wages, either in cash, kind, or as a share of the produce. Household industry refers to an enterprise run by one or more members of a household, either at home or within the village in rural areas, and within the house premises in urban areas. The category of 'other workers' includes individuals engaged in economic activities who do not fall under the previous three classifications (Census of India, 2011a). This comprehensive approach to quantitative data collection provides a robust foundation for understanding demographic and occupational trends in the region over time. The census data on occupational structure and population size of the settlements are provided in Appendix B.

Tourism-related data pertaining to Kargil was obtained from both the District Tourism Department of Kargil and the Tourism Development Authority of LAHDC-Kargil in Ladakh. The influx of both domestic and international tourists was analysed to show the progressive trend in tourist arrivals to Kargil from 1976 to 2024. Additionally, the number of hotels, guesthouses, and travel agencies was examined to gain insights into tourism activities between 2015–16 and 2023–24. The spatial distribution of hotels and guesthouses was also depicted to illustrate the significant growth of tourist facilities from 1974 to 2024. Data on registered transport vehicles were collected from the Regional Transport Office of LAHDC-Kargil and the Ministry of Road Transport and Highways (MoRTHs), India. This data was used to analyse and understand the patterns and trends in registered transport vehicles in Kargil from 2014 to 2022. The data related to tourism and registered transport vehicles are provided in Appendix C.

Climate data was extracted from the NASA Langley Research Center POWER Project, funded through the NASA Earth Science Directorate Applied Science Program. This dataset provides access to meteorological data derived from satellite observations and ground-based measurements. The data was utilised to analyse annual and monthly precipitation and temperature patterns in Kargil. The climate-related data are presented in Appendix E.

3.3 Delineation of the Wards

The thirteen municipal ward boundaries were delineated according to the data given by MC of Kargil using QGIS 3.22 (Fig. 3.2). The settlements outside the municipal wards, particularly the Kurbathang Plateau, were also included in the delineation process. The delineation of ward boundaries was informed by ground truth data and field surveys. The focus of the field surveys was on Wards 1–4, Wards 8, 10, and 13 due to their commercial, administrative, and residential importance. These wards were prioritised because of their significance in various aspects of urban life. Most migrants from rural areas come to Kargil town for education, employment, and job opportunities. In addition to focusing on Kargil town, the field surveys also covered urbanised settlements, including Drass, Sankoo, and Padum, with the overarching aim of comprehensive data collection. This involved engaging with various governmental departments of LAHDC-Kargil to procure essential information and data vital for the study area.

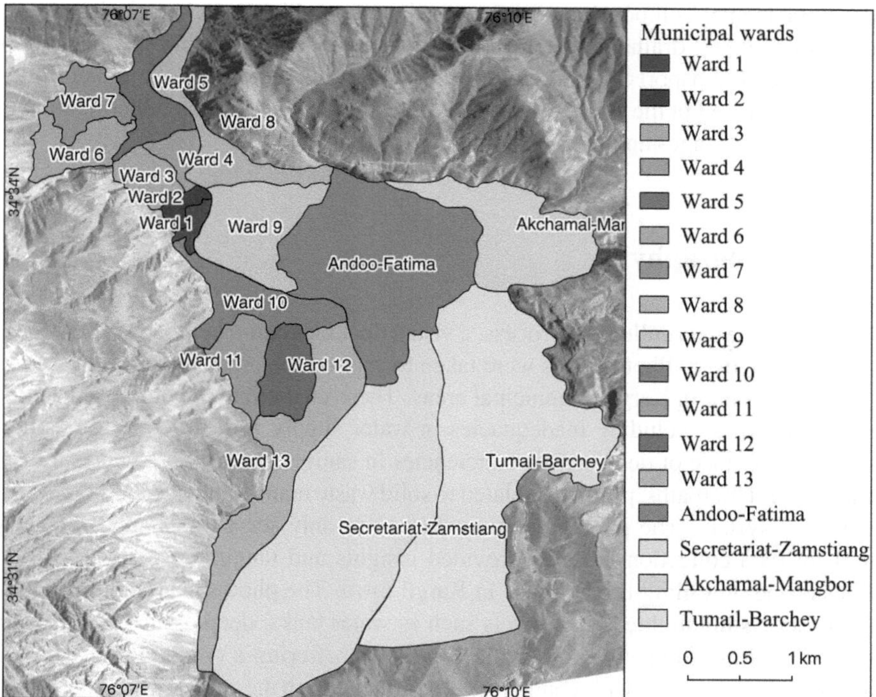

Municipal wards detail: Ward 1 (Changchik-I), Ward 2 (Changchik-II), Ward 3 (Thaskangrong-Lankore), Ward 4 (Daythang-Balti Bazaar), Ward 5 (Dass-Pishu), Ward 6 (Goma Kargil-I), Ward 7 (Goma Kargil-II), Ward 8 (Poyen), Ward 9 (Poyen-Bagh-e-Khomeini), Ward 10 (Baroo-I), Ward 11 (Baroo-II), Ward 12 (Baroo-III), Ward 13 (Baroo-IV). Secretariat-Zamstiang, Andoo-Fatima, Akchamal-Mangbor, and Tumail-Barchey were delineated by the author, as it is not included under Municipal Committee (MC) jurisdiction.

Fig. 3.2 Delineation of municipal wards in Kargil, Ladakh. *Sources* MC Kargil, LAHDC-Kargil; PlanetScope (8 June 2020)

3.4 Urban Mapping

Urban area mapping was carried out using field surveys, remote sensing, and Geographic Information Systems (GIS). Such mapping is essential for effective town planning, urban development, and the overall management of towns and cities. It provides a clear visual representation of urban infrastructure, which helps in identifying issues, analysing spatial patterns, and planning future infrastructure development. The mapping also offers valuable information for planners, policymakers, and decision-makers, supporting evidence-based urban planning. The data collected for this research included land use, infrastructure, and the built environment, which are fundamental components of urban field surveys. Physical characteristics such as building types, drainage system, solid waste management facilities, water supply systems, and transport infrastructure formed the core elements of the urban infrastructure assessed in the study. Analysing these features enabled the identification of the major challenges and problems within the urban area.

3.5 Photography

To support the data collection process, a visual documentation approach was systematically employed. Photographs were taken to capture the urban infrastructure challenges prevalent in Kargil's municipal areas. These challenges encompassed a wide range of issues, including inadequacies in water supply infrastructure, the condition and presence of deep drains, deficiencies in sanitation facilities such as toilets, substandard footpaths, problems related to solid waste management, and a poor transportation system. The photographic evidence not only served as a supplementary tool for data collection but also provided insights and tangible visual representations of the prevailing urban issues in Kargil town. The photos captured during the field surveys highlighted key aspects such as water tanks, deep drains, solid waste, toilets, parking, footpaths, and traffic conditions, offering a comprehensive visual account of the Kargil town's challenges. To further enrich the analysis, repeat photography was utilised as a methodological approach to assess changes in Kargil town's urban structure over time. This technique involves capturing images from identical locations at different time intervals, facilitating a comparative analysis of temporal and spatial transformations. By juxtaposing historical and contemporary images, patterns of urban growth and land-use changes were systematically evaluated. In addition, photographs taken during field visits were used to systematically document and analyse changes in the study area, capturing various dimensions of urban landscape dynamics. These images provided valuable insights into the local environment and socio-economic activities, contributing to a comprehensive understanding of the evolving urban structure and its ongoing transformations.

3.6 Document Collection

Document analysis is a structured process used to review and assess both printed and digital materials (Bowen, 2009). This method involves the systematic evaluation of various documents, including advertisements, books, brochures, manuals, diaries, journals, institutional reports, letters, memoranda, maps, charts, newspapers, and programme proposals, among others. In this study, urban development plans and reports were analysed to gain an in-depth understanding of the existing growth and development in Kargil town. Key documents such as master plans, disaster management plans, tourism development plans, detailed project reports, statistical reports, magazines, local newspapers, and maps were collected from various departments of the LAHDC-Kargil in the UT of Ladakh. Additionally, information on historical development was compiled from written travelogues, gazetteers, journals, books, and other relevant literature on Ladakh.

Furthermore, a detailed literature review was carried out to trace the evolution of town planning practices and provide an overview of urban development trends in Indian cities. This involved examining historical documents and scholarly works to understand the progression of town planning principles and their implementation across various urban centres in India. The review also focused on urban development schemes initiated by local, state, and central governments from the past to the present. This included analysing a wide range of materials such as local government reports, state and policy documents, master plans, academic research, and other relevant sources to gain insights into the trajectory of urban development policies and programmes at different levels of governance. The literature review aimed to develop a thorough understanding of urban development plans, their implementation, and their consequences, focusing specifically on Kargil, Ladakh, by synthesising information from various sources.

3.7 Satellite Imagery

A combined approach using multi-temporal and multi-scale remote sensing data was employed to analyse LULC changes in the Kargil town and the urbanised settlements of Padum, Sankoo, and Drass. Satellite imagery from Corona, the Advanced Spaceborne Thermal Emission and Reflection Radiometer (ASTER), and PlanetScope was utilised for LULC quantification and change detection. The Corona imagery managed by the United States Geological Survey (USGS) refers to declassified reconnaissance satellite images captured by the Corona program, operated by the United States intelligence community during the Cold War era (Cloud, 2001; Fowler, 2013). These images were declassified and transferred to USGS in the 1990s for civilian and scientific use. The Corona imagery managed by USGS includes a vast collection of high-resolution photographs taken from space between 1959 and 1972. These images provide valuable historical records of landscapes, urban areas, and other geographical features during that time period (Dame et al., 2019; Nüsser et al., 2012; Schmidt & Nüsser, 2017).

ASTER is an instrument aboard NASA's Terra satellite, which was launched in 1999. The Terra satellite and its instruments including ASTER are operated by National Aeronautics and Space Administration (NASA), specifically under the oversight of NASA's Earth Observing System program. ASTER collects high-resolution images of the Earth's surface across multiple spectral bands. ASTER satellite imagery is also available in USGS, and it serves as a primary distributor of ASTER data. PlanetScope is operated by Planet (formerly known as Planet Labs), which is a constellation of approximately 130 satellites. These satellites provide high-resolution imagery of the Earth's surface and are part of Planet's larger fleet of satellites designed for Earth observation. PlanetScope satellites have been in operation since the early 2010s. The imagery collected by PlanetScope and ASTER satellites supports multiple applications such as monitoring agriculture, managing urban development, assessing environmental changes, and responding to natural disasters.

The platforms offered by Planet and the USGS were used to access and download Corona, ASTER, and PlanetScope imagery. For Corona and ASTER data, an account was created on the USGS website, imagery was searched by using the Earth Explorer platform, desired imagery was selected, an order was placed, and then the data was downloaded once it became available (https://earthexplorer.usgs.gov/). To download data from PlanetScope, an account was again created on the Planet website, followed by accessing the Planet Explorer platform, searching for imagery of the desired area, selecting preferred imagery based on filters, placing orders, and finally downloading the data once the orders were processed (https://www.planet.com/products/planet-imagery/). ENVI 5.6 and QGIS 3.22 were used for processing, stacking, and mosaicking the satellite images. Corona images were georeferenced by using reference points from Google Earth Pro in QGIS 3.22. Four georeferencing reference points were taken for each dataset of Corona images, and these reference points were identified and linked with ground control points in QGIS 3.22. Shuttle Radar Topography Mission (SRTM) data were used to create a detailed topographic map of the study area in Ladakh, India (Table 3.1).

3.8 Supervised Classification Method

3.8.1 Maximum Likelihood Classification

Remote sensing classification is a sophisticated process that demands careful consideration of multiple influencing factors (Lu & Weng, 2007). Over recent years, various pixel-based classification techniques have been developed to analyse remote sensing imagery, among which the most prominent are the maximum likelihood classifier, support vector machines, and decision tree algorithms (Otukei & Blaschke, 2010; Rimal et al., 2020; Tariq et al., 2023). The classification process generally involves

Table 3.1 Datasets of satellite imagery of the study area in Ladakh

Study area	Satellite	Acquisition date	Scene ID	Spatial resolution (m)	Spectral bands	Source
Kargil	Corona	8 Oct 1965	DS1025-1039DA025_c	2	Pan	USGS
	Corona	17 Jun 1979	DZB1215-500431L009001_b	3	Pan	USGS
	ASTER	4 Sep 2000	AST_L1T_00309042000060203_20150411131129_5580	15 + 30	VNIR/SWIR	USGS
	PlanetScope	8 Jun 2020	20200608_032410_1052_3B_AnalyticMS 20200608_032411_1052_3B_AnalyticMS	3	VIS/NIR	Planet
Padum	Corona	11 Sep 1980	DZB1216-500361L009001_a	10	Pan	USGS
	ASTER	13 Oct 2005	AST_L1T_00310132005054014_20150511131841_53141	15 + 30	VNIR/SWIR	USGS
	PlanetScope	24 Jul 2021	20210724_053504_48_2414_3B_AnalyticMS_	3	VIS/NIR	Planet
Sankoo	Corona	28 Sep 1971	DS1115-2282DF046_b_ DS1115-2282DA053_c_	2	Pan	USGS
	ASTER	18 Jul 2000	AST_L1T_00307182000060217_20150410105406_57728_	15 + 30	VNIR/SWIR	USGS
	PlanetScope	12 Aug 2021	20210812_044922_70_2212_3B_AnalyticMS_	3	VIS/NIR	Planet
Drass	Corona	8 Oct 1965	DS1025-1039DA025_b_	2	Pan	USGS
	ASTER	18 Jul 2000	AST_L1T_00307182000060217_20150410105406_57728_	15 + 30	VNIR/SWIR	USGS
	PlanetScope	18 Jul 2021	20210718_044958_95_2434_3B_AnalyticMS_	3	VIS/NIR	Planet
SRTM		11 Feb 2000	N32_E073 to N32_E083, N33_E073 to N33_E083, N34_E073 to N34_E083, N35_E073 to N35_E083 N36_E073 to N36_E083, N37_E073 to N37_E083	30	C-band	USGS

Sources Hussain et al. (2023); VIS/NIR (Visible/Near-Infrared); VNIR/SWIR (Visible and Near Infrared/Shortwave Infrared); Pan (Panchromatic)

several key steps: defining an appropriate classification framework, selecting training samples, performing image pre-processing, extracting features, choosing a suitable classification method, conducting post-classification adjustments, and evaluating accuracy (Lu & Weng, 2007).

Maximum Likelihood Classification (MLC) is a common technique used in remote sensing, and it is often considered a robust classifier with high precision and accuracy. The classification was performed using the MLC algorithm in ENVI 5.6, which analyses spectral signatures and calculates the probability of each pixel belonging to a specific object class to classify the imagery. The MLC (Liang et al., 2022; Lillesand et al., 2015; Liu et al., 2011; Mohajane et al., 2018; Sun et al., 2013) is a supervised classification method, and it is a commonly used algorithm for classifying LULC (Abou EL-Magd & Kantakumar & Neelamsetti, 2015; Mishra et al., 2020; Mohajane et al., 2018; Tanton, 2003; Weslati et al., 2020) along with other algorithms and methods employed for similar purposes.

A supervised classification was conducted on ASTER and PlanetScope satellite imagery for Kargil town and the urbanised settlements of Padum, Sankoo, and Drass in Ladakh. Training samples were used to generate regions of interest (ROIs) polygons in ENVI 5.6 on the respective satellite images to establish training sites. These training sites were manually digitised on-screen for three land cover classes—built-up area, irrigated land and vegetation, and barren area—based on visual interpretation, supplemented and validated through ground truth data collected during field surveys. The LULC class descriptions and details of satellite images are given in Table 3.2.

Post-classification editing was performed to enhance the accuracy of the results and minimise classification errors (Thakkar et al., 2017). The edit classification image tool was used to reassign classes in classification images that were wrongly assigned into another class. The georeferenced corona images were digitised manually on screen in QGIS 3.22. The digitisation process of LULC was supported by visual interpretation elements (tone, size, pattern, shadow, and association), ground truth knowledge, and Google Earth Pro (historical imagery). It took considerable time and manual effort to improve the classification results of georeferenced Corona images. The methodology used for the satellite image datasets is depicted in Fig. 3.3.

Table 3.2 Description of LULC of the study area in Ladakh

Classes	Class description
Built-up area	Residential, administrative, commercial, army compounds, and other buildings
Irrigated land and vegetation	Cultivated fields, poplars, willows, fruit trees, lucerne (alfalfa), meadows, grasslands, wild herbs, and other trees
Barren area	Barren soils, slope areas without vegetation, and sediment deposits at the bank of rivers

Sources Hussain et al. (2023)

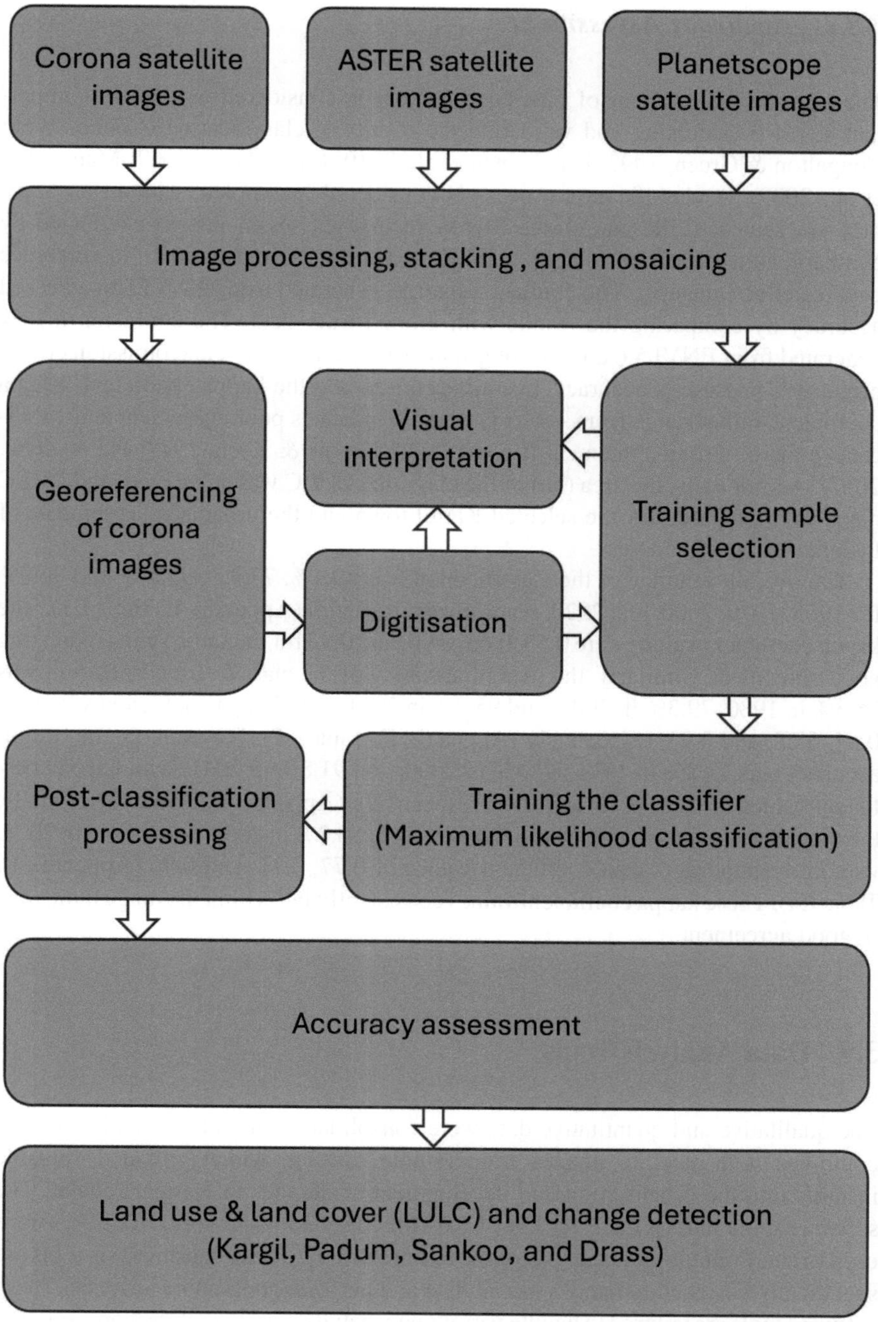

Fig. 3.3 Flowchart methodology for supervised classification used in the study area. Draft by Altaf Hussain

3.8.2 Accuracy Assessment

The accuracy assessment of post-classification is considered as the most important step for evaluating and validating the results of classification (Cohen, 1968; Congalton & Green, 1993; Hurskainen et al., 2019; Lu & Weng, 2007; Manandhar et al., 2009). A classification process is considered incomplete without an accuracy assessment (Lillesand et al., 2015). Accuracy assessment was conducted by comparing the classified images with ground truth ROIs derived from reference data (satellite imagery). The confusion matrix, generated using ENVI 5.6, assessed accuracy by comparing the results with ground truth data. The confusion matrix generated from ENVI 5.6 consists of ground truth, commission and omission errors, producer's and user's accuracy, overall accuracy, and the kappa coefficient. Kappa coefficient values range from −1 to 1, where 0 indicates poor agreement and 1 indicates good or strong agreement (Foody, 2002; Landis & Koch, 1977; Pal & Ziaul, 2017). Additionally, the area (ha) of the classified LULC was calculated, and LULC maps were prepared for the selected Kargil town and the urbanised settlements of Padum, Sankoo, and Drass.

The overall accuracy of the classification was 69.1%, 77.3%, 93.5%, and 96.9% for 1965, 1979, 2000, and 2020, respectively, in Kargil (Appendix E, Table E1). The kappa coefficient values were 0.53, 0.65, 0.90, and 0.95 for the same years, indicating good agreement. Similarly, the overall accuracy of the classification in Padum was 76.3% in 1980, 79.3% in 2005, and 95.8% in 2021, with kappa coefficient values of 0.64, 0.68, and 0.93, respectively (Appendix E, Table E2). For Sankoo, the overall accuracy was 73.2% in 1971, 82.4% in 2000, and 91.8% in 2021, with kappa coefficient values of 0.55, 0.71, and 0.87, respectively (Appendix E, Table E3). Finally, the overall accuracy in Drass was 84.9% in 1965, 80.8% in 2000, and 97.9% in 2021, with corresponding kappa coefficient values of 0.77, 0.71, and 0.96 (Appendix E, Table E4). These kappa coefficient values consistently fall within the range indicative of good agreement.

3.9 Data Analysis Tools

The qualitative and quantitative data were consolidated, interpreted, and analysed, combined with satellite images from remote sensing, which provided valuable insights into the urbanisation and development challenges in Kargil, Ladakh. The software tools like QGIS 3.22, ENVI 5.6, ArcGIS, and Microsoft Office 365 were used to carry out the data analysis of the research study. These data analysis tools are specifically designed to handle spatial data and perform geographic analysis. These software tools are tailored to handle various geospatial datasets such as maps, satellite imagery, satellite navigation, and spatial data. These tools are essential for processing, visualising, and interpreting geographic information to understand spatial patterns, trends, and relationships. The diagrams, graphs, and thematic maps were prepared by using these software tools. For instance, satellite imagery from 11 February 2000 (SRTM) and OpenStreetMap (OSM) were employed to map the study area in Kargil, Ladakh. The historical development was showcased from Tibetan rule to the contemporary era through a chronological timeline. Using digitisation and overlaying techniques, the historical trade networks of Ladakh during the nineteenth and early twentieth centuries were depicted. Similarly, the road network was represented to illustrate the development of road connectivity in the Ladakh region after the partition of India. The advancement of open-source GIS software, especially QGIS, has enabled users to access data through freely available plugins. This comprehensive suite of tools offers researchers and practitioners an effective platform to perform various GIS tasks, including geoprocessing, statistical analysis, and cartographic design (Garnett & Kanaroglou, 2016).

References

Abou EL-Magd, I., & Tanton, T. W. (2003). Improvements in land use mapping for irrigated agriculture from satellite sensor data using a multi-stage maximum likelihood classification. *International Journal of Remote Sensing, 24*(21), 4197–4206. https://doi.org/10.1080/0143116031000139791

Baker, L. (2006). Observation: A complex research method. *Library Trends, 55*(1), 171–189. https://doi.org/10.1353/lib.2006.0045

Bloomberg, L. D., & Volpe, M. (2019). *Completing your qualitative dissertation: A road map from beginning to end* (4th ed.). Sage.

Bowen, G. A. (2009). Document analysis as a qualitative research method. *Qualitative Research Journal, 9*(2), 27–40. https://doi.org/10.3316/QRJ0902027

Census of India. (1961). *Primary census abstract*. Office of the Registrar General & Census Commissioner, India. https://censusindia.gov.in

Census of India. (1971). *Primary census abstract*. Office of the Registrar General & Census Commissioner, India. https://censusindia.gov.in

Census of India. (1981). *Primary census abstract*. Office of the Registrar General & Census Commissioner, India. https://censusindia.gov.in/

Census of India. (2001). *Primary census abstract*. Office of the Registrar General & Census Commissioner, India. https://censusindia.gov.in/

Census of India. (2011a). *District census handbook Kargil: Village and town directory, series-02, part XII-A*. Directorate of Census Operations, Jammu & Kashmir.

Census of India. (2011b). *District census handbook Kargil: Village and town wise, series-02, part XXI-B*. Directorate of Census Operations, Jammu & Kashmir.

Census of India. (2011c). *Primary census abstract*. Office of the Registrar General & Census Commissioner, India. https://censusindia.gov.in/

Cloud, J. (2001). Imaging the world in a barrel: CORONA and the clandestine convergence of the earth sciences. *Social Studies of Science, 31*(2), 231–251. https://doi.org/10.1177/0306312701031002005

Cohen, J. (1968). Weighted kappa: Nominal scale agreement provision for scaled disagreement or partial credit. *Psychological Bulletin, 70*(4), 213–220. https://doi.org/10.1037/h0026256

Congalton, R. G., & Green, K. (1993). A practical look at the sources of confusion in error matrix generation. *Photogrammetric Engineering and Remote Sensing, 59*(5), 641–644.

Creswell, J. W. (1999). Mixed-method research. In *Handbook of educational policy* (pp. 455–472). Elsevier. https://doi.org/10.1016/B978-012174698-8/50045-X

Creswell, J. W., & Clark, V. L. P. (2017). *Designing and conducting mixed methods research* (2nd ed.). Sage Publications.

Dame, J., Schmidt, S., Müller, J., & Nüsser, M. (2019). Urbanisation and socio-ecological challenges in high mountain towns: Insights from Leh (Ladakh), India. *Landscape and Urban Planning, 189*, 189–199. https://doi.org/10.1016/j.landurbplan.2019.04.017

District Statistical & Evaluation Office Kargil. (2012). *District Kargil at a Glance.* Directorate of Economics & Statistics, Planning Development & Monitoring Department.

District Statistical & Evaluation Office Kargil. (2016). *Statistical handbook.* Directorate of Economics & Statistics, Planning Development & Monitoring Department.

District Statistical & Evaluation Office Kargil. (2018). *Statistical handbook.* Directorate of Economics & Statistics, Planning Development & Monitoring Department.

District Statistical & Evaluation Office Kargil. (2019). *Statistical handbook.* Directorate of Economics & Statistics, Planning Development & Monitoring Department. https://cdn.s3waas.gov.in/s341ae36ecb9b3eee609d05b90c14222fb/uploads/2020/03/2020031274.pdf

District Statistical & Evaluation Office Kargil. (2020a). *Statistical handbook.* Directorate of Economics & Statistics, Planning Development & Monitoring Department. https://cdn.s3waas.gov.in/s341ae36ecb9b3eee609d05b90c14222fb/uploads/2021/07/2021071951.pdf

District Statistical & Evaluation Office Kargil. (2020b). *Village amenity directory.* Directorate of Economics & Statistics, Planning Development & Monitoring Department.

District Statistical & Evaluation Office Kargil. (2021). *Statistical Handbook.* Directorate of Economics & Statistics, Planning Development & Monitoring Department.

District Statistical & Evaluation Office Kargil. (2022). *Statistical handbook.* Directorate of Economics & Statistics, Planning Development & Monitoring Department.

District Statistical & Evaluation Office Kargil. (2023). *Statistical handbook.* Directorate of Economics & Statistics, Planning Development & Monitoring Department.

District Statistical & Evaluation Office Kargil. (2024). *Statistical handbook.* Directorate of Economics & Statistics, Planning Development & Monitoring Department.

Foody, G. M. (2002). Status of land cover classification accuracy assessment. *Remote Sensing of Environment, 80*(1), 185–201. https://doi.org/10.1016/S0034-4257(01)00295-4

Fossey, E., Harvey, C., Mcdermott, F., & Davidson, L. (2002). Understanding and evaluating qualitative research. *Australian & New Zealand Journal of Psychiatry, 36*(6), 717–732. https://doi.org/10.1046/j.1440-1614.2002.01100.x

Fowler, M. J. F. (2013). Declassified intelligence satellite photographs. In W. S. Hanson, & I. A. Oltean (Eds.), *Archaeology from historical aerial and satellite archives* (pp. 47–66). Springer New York. https://doi.org/10.1007/978-1-4614-4505-0_4

Garnett, R., & Kanaroglou, P. (2016). Qualitative GIS: An open framework using SpatiaLite and open source GIS. *Transactions in GIS, 20*(1), 144–159. https://doi.org/10.1111/tgis.12163

Hurskainen, P., Adhikari, H., Siljander, M., Pellikka, P. K. E., & Hemp, A. (2019). Auxiliary datasets improve accuracy of object-based land use/land cover classification in heterogeneous savanna landscapes. *Remote Sensing of Environment, 233*, 111354. https://doi.org/10.1016/j.rse.2019.111354

Hussain, A., Schmidt, S., & Nüsser, M. (2023). Dynamics of mountain urbanisation: Evidence from the Trans-Himalayan town of Kargil, Ladakh, India. *Land, 12*(4), 920. https://doi.org/10.3390/land12040920

Johnson, R. B., Onwuegbuzie, A. J., & Turner, L. A. (2007). Toward a definition of mixed methods research. *Journal of Mixed Methods Research, 1*(2), 112–133. https://doi.org/10.1177/1558689806298224

Kantakumar, L. N., & Neelamsetti, P. (2015). Multi-temporal land use classification using hybrid approach. *The Egyptian Journal of Remote Sensing and Space Science, 18*(2), 289–295. https://doi.org/10.1016/j.ejrs.2015.09.003

Landis, J. R., & Koch, G. G. (1977). The measurement of observer agreement for categorical data. *Biometrics, 33*(1), 159. https://doi.org/10.2307/2529310

Liang, F., Zhang, X., Li, H., Yu, H., Lin, Q., Jiang, M., & Zhang, J. (2022). Land use classification based on maximum likelihood method. In J.-S. Pan, V. E. Balas, & C.-M. Chen (Eds.), *Advances in intelligent data analysis and applications* (Vol. 253, pp. 133–139). Springer Singapore. https://doi.org/10.1007/978-981-16-5036-9_15

Lillesand, T. M., Kiefer, R. W., & Chipman, J. W. (2015). *Remote sensing and image interpretation* (7th ed.). John Wiley & Sons, Inc.

Liu, K., Shi, W., & Zhang, H. (2011). A fuzzy topology-based maximum likelihood classification. *ISPRS Journal of Photogrammetry and Remote Sensing, 66*(1), 103–114. https://doi.org/10.1016/j.isprsjprs.2010.09.007

Lu, D., & Weng, Q. (2007). A survey of image classification methods and techniques for improving classification performance. *International Journal of Remote Sensing, 28*(5), 823–870. https://doi.org/10.1080/01431160600746456

Manandhar, R., Odeh, I., & Ancev, T. (2009). Improving the accuracy of land use and land cover classification of Landsat data using post-classification enhancement. *Remote Sensing, 1*(3), 330–344. https://doi.org/10.3390/rs1030330

Merriam, S. B., & Tisdell, E. J. (2015). *Qualitative research: A guide to design and implementation* (4th ed.). John Wiley & Sons.

Mishra, P. K., Rai, A., & Rai, S. C. (2020). Land use and land cover change detection using geospatial techniques in the Sikkim Himalaya, India. *The Egyptian Journal of Remote Sensing and Space Science, 23*(2), 133–143. https://doi.org/10.1016/j.ejrs.2019.02.001

Mohajane, M., Essahlaoui, A., Oudija, F., El Hafyani, M., Hmaidi, A. E., El Ouali, A., Randazzo, G., & Teodoro, A. C. (2018). Land use/land cover (LULC) using Landsat data series (MSS, TM, ETM+ and OLI) in Azrou forest, in the Central Middle Atlas of Morocco. *Environments, 5*(12), 131. https://doi.org/10.3390/environments5120131

Morgan, D. L. (2013). *Integrating qualitative and quantitative methods: A pragmatic approach.* Sage Publications.

Nüsser, M., Schmidt, S., & Dame, J. (2012). Irrigation and development in the Upper Indus Basin: Characteristics and recent changes of a socio-hydrological system in Central Ladakh, India. *Mountain Research and Development, 32*(1), 51–61. https://doi.org/10.1659/MRD-JOURNAL-D-11-00091.1

Otukei, J. R., & Blaschke, T. (2010). Land cover change assessment using decision trees, support vector machines and maximum likelihood classification algorithms. *International Journal of Applied Earth Observation and Geoinformation, 12*, S27–S31. https://doi.org/10.1016/j.jag.2009.11.002

Pal, S., & Ziaul, Sk. (2017). Detection of land use and land cover change and land surface temperature in English Bazar urban centre. *The Egyptian Journal of Remote Sensing and Space Science, 20*(1), 125–145. https://doi.org/10.1016/j.ejrs.2016.11.003

Patton, M. Q. (2005). Qualitative research. In B. S. Everitt, & D. C. Howell (Eds.), *Encyclopedia of statistics in behavioral science* (1st ed.). Wiley. https://doi.org/10.1002/0470013192.bsa514

Philip, L. J. (1998). Combining quantitative and qualitative approaches to social research in human geography—An impossible mixture? *Environment and Planning A: Economy and Space, 30*(2), 261–276. https://doi.org/10.1068/a300261

Rimal, B., Rijal, S., & Kunwar, R. (2020). Comparing support vector machines and maximum likelihood classifiers for mapping of urbanization. *Journal of the Indian Society of Remote Sensing, 48*(1), 71–79. https://doi.org/10.1007/s12524-019-01056-9

Schmidt, S., & Nüsser, M. (2017). Changes of high altitude glaciers in the Trans-Himalaya of Ladakh over the past five decades (1969–2016). *Geosciences, 7*(2), 27. https://doi.org/10.3390/geosciences7020027

Silverman, R. M., & Patterson, K. L. (2022). *Qualitative research methods for community development* (2nd ed.). Routledge, Taylor & Francis Group.

Sun, J., Yang, J., Zhang, C., Yun, W., & Qu, J. (2013). Automatic remotely sensed image classification in a grid environment based on the maximum likelihood method. *Mathematical and Computer Modelling, 58*(3–4), 573–581. https://doi.org/10.1016/j.mcm.2011.10.063

Tariq, A., Jiango, Y., Li, Q., Gao, J., Lu, L., Soufan, W., Almutairi, K. F., & Habib-ur-Rahman, M. (2023). Modelling, mapping and monitoring of forest cover changes, using support vector machine, kernel logistic regression and Naive Bayes tree models with optical remote sensing data. *Heliyon, 9*(2), e13212. https://doi.org/10.1016/j.heliyon.2023.e13212

Teddlie, C., & Tashakkori, A. (2009). *Foundations of mixed methods research: Integrating quantitative and qualitative approaches in the social and behavioral sciences.* SAGE Publications, Inc.

Thakkar, A. K., Desai, V. R., Patel, A., & Potdar, M. B. (2017). Post-classification corrections in improving the classification of land use/land cover of arid region using RS and GIS: The case of Arjuni watershed, Gujarat, India. *The Egyptian Journal of Remote Sensing and Space Science, 20*(1), 79–89.

Watson, R. (2015). Quantitative research. *Nursing Standard, 29*(31), 44–48. https://doi.org/10.7748/ns.29.31.44.e8681

Weslati, O., Bouaziz, S., & Serbaji, M. M. (2020). Mapping and monitoring land use and land cover changes in Mellegue watershed using remote sensing and GIS. *Arabian Journal of Geosciences, 13*(14), 1–19. https://doi.org/10.1007/s12517-020-05664-5

Chapter 4
Historical Development of Ladakh

Abstract Ladakh, situated in a geopolitically sensitive border region in northern India, has undergone significant administrative changes due to evolving geopolitical circumstances and governance reforms. This chapter examines the historical development of Ladakh, beginning with the disintegration of the Tibetan monarchy in 842 A.D., continuing through Dogra rule, the Colonial period, the Trans-Karakoram trade, the emergence of Islam, and into the post-independence era. Additionally, it provides an overview of the contemporary social, religious, and political developments in Kargil, Ladakh. The chapter explores how these historical phases have influenced the region's cultural, political, and economic landscape, including the integration of diverse communities and the impact of external influences on local traditions. A thorough understanding of this historical context is crucial for analysing the contemporary situation in Ladakh, India, particularly in light of recent geopolitical tensions and administrative reorganisation.

Keywords Ladakh kingdom · Dogra rule · Colonial period · Trans-Karakoram trade · Traders · Kingdom · Islam · Tibet · Kargil · Leh · Skardu · Ladakh · Kashmir · LAHDC-Kargil · India

4.1 Ladakh Kingdom

Ladakh was once under the Tibetan Empire, which began to disintegrate following the assassination of King Langdarma by the monk Palgyi Dorje in 742 A.D., and likely continued until the fall of the central Tibetan monarchy around 842 A.D. (Bray, 1991; Petech, 1977; Rizvi, 2012). In the early tenth century, Kyide Nyimagon consolidated power across the broader Western Tibetan region. Following his death around 950 A.D., his extensive kingdom was divided among his three sons (Francke, 1926; Petech, 1977; Rizvi, 2012). Lhachen Palgyigon inherited Ladakh, while Trashigon took control of Guge and Purang, and Detsukgon was granted Zanskar and Spiti (Bray, 2005; Petech, 1977; Rizvi, 2012). Palgyigon founded the kingdom of Ladakh,

A. Hussain, *Urbanisation and Development Challenges in Kargil, Ladakh*, Advances in Asian Human-Environmental Research, https://doi.org/10.1007/978-3-032-05667-2_4

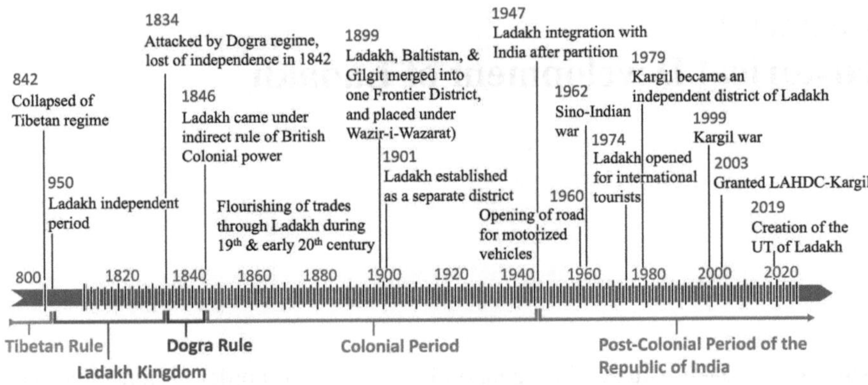

Fig. 4.1 Historical context of the development of Kargil in Ladakh, India. Draft by Altaf Hussain

and all succeeding kings of Ladakh[1] claimed descent from him (Bray, 2005; Petech, 1977; Rizvi, 2012). Hence, Ladakh continued to be an independent kingdom between 950 A.D and 1834 (Bray, 2005) (Fig. 4.1). Ladakh was often at war with Central Asian Mughal rulers, and Mirza Haider Dughlat attacked the region in 1532, 1545, and 1548 (Bray, 2005; Rizvi, 2012).

The Namgyal dynasty took over the Ladakhi Kingdom from the early sixteenth century onward. Around 1600 A.D., Muslim rulers attempted to extend Islamic influence over Ladakh, and Jamyang Namgyal (1595–1616) became involved in a dispute between the chiefs of the Sultans of Chiktan and Kartse, as they had just embraced Islam (Rizvi, 2012). The king assembled an army to aid of the Sultan of Chiktan (Tsering Malik), but the whole situation was an elaborate plot orchestrated by a cabal of Muslim rulers. When he crossed Fatu La, he was challenged by the troops of Ali Mir (Ali Sher Khan Anchan) of the Maqpon dynasty, a prominent figure in the history of Baltistan (Petech, 1977; Rizvi, 2012). The Balti forces defeated the Ladakhi army, and Jamyang Namgyal was taken prisoner in Skardu by Ali Mir during the early seventeenth century (Petech, 1977; Rizvi, 2012; Sheikh, 2014). To restore his throne, Ali Mir imposed peace terms on Jamyang Namgyal, which stipulated that his new Balti bride (Gyal Khatun) was to be made the first queen (Petech, 1977; Rizvi, 2012; Sheikh, 2014). The new queen (Gyal Khatun) bore the king (Gyalpo) two sons, Sengge Namgyal and Norbu Namgyal (Petech, 1977).

Sengge Namgyal (1616–1642) was one of the greatest and best-known Ladakhi kings (Petech, 1977). In 1630, he annexed all of Western Tibet and enlarged the Ladakhi empire to its greatest extent (Bray, 2005; Rizvi, 2012). However, when he invaded and conquered Purig (probably in 1635), Adam Khan of Skardo called for the aid of the Mughul forces under the Governor of Kashmir, and a conflict occurred at Kharbu in Purig, where the Ladakhis were decisively defeated (Petech, 1939).

[1] The Ladakhi prince Gyalpo Rinchen extended his rule to Kashmir and accepted Islam. He ruled under the name Sultan Sadar-ud-din (Rinchan Shah) from 1324 to 1327 and was the first Muslim king of Kashmir (Rizvi, 2012; Schettler & Schettler, 1989).

During Sengge Namgyal's reign, the first European merchant, Diogo d'Almeida (a Portuguese), visited Ladakh and described it as a rich country (Bray, 2005; Petech, 1939; Rizvi, 2012). Sengge also built the largest and most famous monasteries such as those at Hanle and Hemis; and constructed the nine-storey royal palace in Leh (Bray, 2005; Petech, 1977; Rizvi, 2012). He was succeeded by his eldest son, Deldan Namgyal (1642–1675). Deldan Namgyal aimed to restore the lost prestige of the Ladakhi army, bring the rulers of Purig under his dominion, and avenge his family's honour against the Baltis (Francke, 1995). The first territory they invaded was Kharbu, ruled by Khri Sultan of Kartse, an independent king (Gyalpo) who ruled from Kharbu to Drass (Francke, 1995). The Ladakhi army attacked the territory of other Purig rulers, and the Purig Sultan of Chiktan, along with his strongest castles (Chiktan and Shaker), were taken. Once the threat from that side was neutralised, the army marched against the three strongest castles of the Khri Sultan. First, they attacked Sodh, followed by Sumtang and Kartse, and captured them all. The Khri Sultan was captured and transferred to Leh (Francke, 1995). Following the battle of Kharbu, King Sengge Namgyal pledged to pay tribute to the Mughal emperor; however, the tribute was never delivered, and Ladakh continued to function as an independent state (Petech, 1977). This issue was brought to Aurangzeb's attention during his visit to Kashmir (Khachul) in 1663.

On this occasion, Deldan Namgyal reaffirmed his father's commitments to pay tribute and show loyalty, pledging to construct a mosque, recite the Khutba, and mint coins bearing the emperor's name (Rizvi, 2012). Deldan was compelled to adopt Islam, taking the name Aqibat Mahmud Khan, the title by which the Mughal and Kashmir authorities thereafter referred to the kings of Ladakh (Petech, 1977; Rizvi, 2012).

Delegs Namgyal (1675–1705), son of Deldan Namgyal, proved to be an inadequate successor to his father and grandfather. During his reign, the great Ladakhi empire crumbled and collapsed (Petech, 1977). In 1681, hostilities commenced between Ladakh and the Lhasa government, and the Ladakhi empire proved unable to withstand the Tibetan forces. As a result, it sought assistance from Ibrahim Khan, the Governor of Kashmir, who was himself under the authority of the Mughal emperor (Bray, 1991; Petech, 1977). The Treaty of Tingmosgang (1684) resolved the contested relations between Tibet and Ladakh and brought an end to the four-year-long conflict between the two regions (Bray, 2012; Petech, 1977; Rizvi, 2004). The borders were demarcated, and trade sovereignties were defined under the forceful influence of the Mughal Empire (Bray, 2012; Petech, 1977). Kashmiri merchants gained a monopoly over the pashmina trade, while Ladakh relinquished Guge, Rudok, and Purang to the Lhasa government. Additionally, Ladakh agreed to send a triennial mission (Lopchak) of homage and tribute from the king (Gyalpo) of Ladakh to the Tibetan (Lhasa) religious establishment on the occasion of the New Year (Bray, 2005, 2012; Petech, 1977; Rizvi, 2004). The Tibetan caravan, which arrived once a year from Lhasa to Leh carrying brick-tea and various other goods, was known as the Chapa (Bray, 1991; Rizvi, 2004; Warikoo, 1989). The Chapa caravan was granted free transportation and lodging throughout its journey and stay in Ladakh, and Ladakhi traders were offered equivalent arrangements during their travels in Tibet (Rizvi,

2004; Warikoo, 1989). The Treaty of Tingmosgang (1684) also led to an agreement between Ladakh and the Mughals, whereby all fine wool (bal) imported from Tibet was to be exclusively delivered to Kashmir (Khachul) (Warikoo, 1990). Shawl-wool, as the principal trade commodity between Ladakh and Tibet, emerged as a profitable source of revenue for Ladakh's ruling elite, as well as for the limited number of Kashmiri and Arghon[2] traders who enjoyed exclusive access to Tibet (Warikoo, 1990).

Delegs Namgyal was succeeded by Nyima Namgyal (1705–1734), under whose reign the Jesuit missionary Father Ippolito Desideri visited Ladakh in 1715 (Petech, 1939). Nyima Namgyal held a distinguished place among Ladakhi rulers due to his contributions to justice and the judicial system (Petech, 1977). Culturally, he promoted printing activities—printing blocks for devotional texts were produced, and numerous religious structures such as images, mani walls, and prayer wheels were constructed (Petech, 1977). King Nyima married twice; his first wife passed away after the birth of their son, Lhachen Deskyong Namgyal. He later married Zizi Khatun, the granddaughter of Hatim Khan and niece of Daulat Khan of Khaplu, with whom he had a son, Trashi Namgyal, and a daughter, Trashi Wangmo (Francke, 1995; Petech, 1977). Nyima Namgyal was followed in 1734 by Deskyong Namgyal (1734–1750), whose successors were Puntsog Namgyal (1750–1765), Tsewang Namgyal (1765–1780), and Tsestan Namgyal (1780–1790) (Petech, 1939). Tsepal Namgyal (brother of Tsestan Namgyal) enjoyed an unusually long reign. He ascended the throne by 1792 and remained in power until his deposition by the Dogras in 1834 (Petech, 1939).

4.2 Dogra Conquest

The Ladakh kingdom in the Trans-Himalayan region was brutally attacked by the Dogra forces under Raja Gulab Singh of Jammu during the Dogra rule. In 1834, Raja Gulab Singh decided to conquer Ladakh and raised an army of about 5,000 soldiers. He appointed his most trusted lieutenant, Zorawar Singh, to lead the force and assigned him the mission of conquering Ladakh (Petech, 1977) (Fig. 4.1). Ladakh was an independent kingdom before it was invaded by Zorawar Singh in August 1834 (Cunningham, 1854) and lost its independence in 1842 (Bray, 1991). In July 1834, Zorawar Singh departed from Kishtwar, crossed the Bhot Kol pass (Suru Valley), and entered Purig (Kargil) with his forces (Petech, 1977). On 16 August 1834, Ladakhi forces of approximately 5,000, led by Stog Lonpo Dorje Namgyal (or Boti leader Mangal), fought a battle against the invaders of Dogras army at Sankoo, but were eventually dislodged and defeated (Cunningham, 1854; Francke, 1926; Petech, 1977). The Dogra forces then marched to Suru, where they halted for some

[2] Many Muslim immigrants from Kashmir and Central Asia married with local Ladakhi women, and their descendants belonging to a hybrid class are known as Arghon. This mixed race of Sunni Muslim mostly resides in Leh.

time and built a small fort (khar) (Datta, 1973). Zorawar Singh advanced to Pashkyum and was opposed by the Ladakhis at the bridge of Pashkyum (Cunningham, 1854; Gazetteer of Kashmir and Ladák, 1890). The Dogras managed to cross the bridge as the Ladakhis retreated and broke down, and the local chief abandoned Pashkyum and fled to the fort (khar) of Sodh (Francke, 1926). The attackers advanced towards Sodh and began cannonading the fort (khar), which had been heavily fortified by the Ladakhi forces (Datta, 1973). Salam Khan, the Kiladar of Sodh, put up a valiant defence, and despite ten days of continuous firing, the Dogras failed to make any gains, suffering the loss of forty soldiers, either killed or wounded (Cunningham, 1854; Datta, 1973). Zorawar Singh, who had stayed back at Pashkyum, then ordered Mehta Basti Ram to lead a contingent of 500 troops in a vigorous assault on the place, and they ultimately took possession of it (Cunningham, 1854). They captured the chief, and a total of 6,000 Ladakhi troops were taken prisoner at Pashkyum and Sodh (Francke, 1926). After the negotiation and disagreement over the tribute with the Dogras, substantial reinforcements were dispatched for Purig under the command of Leh Nosgrub Stanzin and Nubra Dorje Stanzin. In April 1835, another battle was fought at Lankerchey, which ended in the utter defeat of the Ladakhis (Petech, 1977). Although this battle was decisive, Banka Kahlon attacked the rear of the invaders and later fled to Baltistan. After this triumph, the Dogras were once again able to march towards Pashkyum, making use of their prisoners to carry their baggage. They marched from Shergol to Mulbekh, then continued from Bodh Kharbu to Lamayuru, where an envoy met them carrying a letter from Sultan Akabat Mahmud Khan, the King of Ladakh, requesting peace (Cunningham, 1854; Francke, 1926). An agreement was negotiated and signed between the Ladakhi king (Gyalpo) and the Dogras (led by Zorawar Singh), which stipulated that the Gyalpo of Ladakh would pay a yearly tribute of ₹20,000 along with war reparations of ₹50,000 (Cunningham, 1854; Francke, 1926; Petech, 1977) (Fig. 4.2).

4.3 Colonial Period

The Treaty of Amritsar, signed in March 1846 between the East India Company and Dogra ruler Maharaja Gulab Singh, led to the formation of the princely state of Jammu and Kashmir (Huttenback, 1973). Ladakh and Baltistan were governed by a senior official of Kashmir known as the Wazir-i-Wazarat, based in Leh. Under his authority were two tehsildars—one stationed in Kargil and the other in Skardu. An English official, stationed at Leh, represented the British Government in these two districts and worked under the supervision of the Resident of Kashmir (De Filippi, 1912). Ladakh was incorporated into the princely state and was brought under the indirect rule of the British colonial power in 1846 (Bray, 1991). The British Joint Commissioner, in collaboration with the Kashmiri Wazir (Governor), supervised the Central Asian trade passing through Ladakh. However, the Wazir held primary responsibility for tax collection and local administrative matters in the region (Bray, 1991). As a result, the British were never able to establish direct rule over Ladakh.

Fig. 4.2 Dogra fort in Leh, Ladakh. *Sources* Helmut De Terra, photograph taken between 1932 and 1933, American Geographical Society Library, University of Wisconsin-Milwaukee Libraries

The Treaty of Amritsar, signed in 1846, formally recognised Gulab Singh as the ruler not only of the princely state of Jammu and Kashmir but also of Ladakh and Baltistan. In 1899, Ladakh, Baltistan, and Gilgit were consolidated into a single administrative unit known as the 'Frontier District', which was placed under the authority of the Wazir-i-Wazarat (Dani, 1991; Warikoo, 2014). In 1901, two years later, Ladakh was established as a separate district, incorporating the tehsils of Skardu, Kargil, and Leh (Dani, 1991; Warikoo, 2014).

This administrative unit, known as the Ladakh Wazarat, encompassed the territories of Leh, Kargil, and Skardu tehsils, along with the Zanskar region, which had previously been part of the Kishtwar tehsil in the Jammu Province (Khan, 1902). The headquarters of the Wazarat were located in Leh during the summer and in Skardu during the winter (Sheikh, 2014).

Moreover, Eduard Pagell and Wilhelm Heyde were the first Moravian missionaries[3] to arrive in Ladakh. In the summer of 1855, they conducted a reconnaissance journey through Zanskar to Leh and the Tibetan frontier (Bray, 1983). In May 1856, the two Moravian pioneers founded the first Himalayan mission station in Kyelang, Lahaul, and the mission opened new stations at Poo on the upper Sutlej in 1865 and at Leh in 1885 (Bray, 1983, 2008). In 1857, Heinrich August Jäschke, the new superintendent of the mission, arrived and began his linguistic works. His major contributions included the publication of a *Tibetan-English dictionary* in 1881 and the translation of the New Testament into Tibetan (Bray, 1983, 2008). August Hermann Francke arrived in Leh, the capital of Ladakh, on 8 June 1896. He was the first missionary to establish a new mission station at Khaltse in 1899. Franke was among the most distinguished missionary scholars of the Moravian Church (Herrnhuter Brüdergemeine) to have worked in the Himalayan region. His key contributions included linguistic research on the Tibetan language—particularly the Ladakhi dialect—as well as groundbreaking work on Ladakh's history, folk tales, songs, and rock inscriptions (Bray, 1983, 2008). One of the early converts, Joseph Gergan, was the son of a Tibetan scholar who had married and settled in Nubra. Fluent in classical Tibetan, Gergan translated 29 books of the Old Testament (Rizvi, 2012).

4.4 Trans-Karakoram Trade

Ladakh served as a vibrant trading hub along the Silk Route connecting Central and South Asia, with its strategic geographic position playing a crucial role in facilitating this trade (Fewkes, 2008). The region was also a centre of a network of trade routes until 1947, which connected Skardu, Srinagar, Hoshiarpur, Lahul, Kulu, Lhasa, Baltistan, Rudok, and Gartok (Rizvi, 1985, 2004). The majority of trade (tsonglen) exchanges with the Central Asian centres of Kashgar, Yarkand, and Khotan were carried out through Ladakh (Warikoo, 1989). The cosmopolitan town of Leh was a hub of trading networks and used to offer accessibility for traders on different trading routes (Fewkes, 2008; Rizvi, 2004). Leh and Kargil, the primary urban centres of Ladakh, functioned as bustling bazaars and key focal points of intense commercial activity, where commodities were traded, taxed, and transported (Fewkes, 2006, 2008). In the early 1870s, exports from Leh to India (Kullu, Punjab, and Kashmir) nearly doubled from ₹169,499 in 1872 to ₹329,036 in 1873, driven primarily by increased trade in silks, pashm, charas, gold, silver, and horses (Aitchison, 1874). During this period, there was a notable increase in the export trade (tsonglen) of shawl-wool (pashm) from Leh to Kashmir, which surged by ₹55,661 and more than

[3] The earliest Christian missionaries to arrive in Ladakh were Jesuits Francisco de Azevedo, John de Oliveira, and Ippolito Desideri, who visited the region during the seventeenth and eighteenth centuries (Rizvi, 2004).

tripled since 1872. Lhasa tea exports also witnessed a rise of ₹36,900 (Aitchison, 1874). The value of the trade (including both import and export) rose from ₹3,400,822 in 1885 to ₹6,015,321 in 1894–95 in Ladakh (Adair & Godfrey, 1899).

The Central Asian exports of commodities to Ladakh included gold, shawl-wool, silver, tea (cha), cannabis hemp (charas), carpets, cotton cloths, raw silk, spices, and ponies (Fewkes, 2008; Rizvi, 2004; Warikoo, 1989). The Yarkandi and Andijani merchants brought these goods into Ladakh, where they engaged in exchanges with other trading communities (Rizvi, 2004; Warikoo, 1989). The trade (tsonglen) networks in Ladakh involved diverse communities, including traders from British Indian Punjab—such as the Gaddis or Lalas—and those from Chinese Central Asia, including the Yarkandis, Hors, and Tungan Chinese (Fewkes, 2008). The Ladakhi Arghon traders occupied an elite position within their community, and commercial goods played a significant role in shaping their social relationships both within Ladakh and beyond, particularly through their interactions with trading partners from Chinese Central Asia and British Indian Punjab (Fewkes, 2008). Bahauddin Khan and Munshi Aziz Bhat[4] were the most prominent traders, who carried out long-distance trades and commanded large businesses and mercantile goods (Fewkes, 2008) (Fig. 4.3).

The development of Kashmir as the leading centre for shawl production during the medieval period brought a considerable number of Kashmiri traders into Ladakh and its surrounding areas. A Kashmiri community involved in the collection and shipment of shawl-wool to Kashmir settled in Ladakh. They bartered raw wool (bal) for finished shawl goods, which were subsequently traded across Central Asian markets (Warikoo, 1990). Moreover, the trade (tsonglen) connection also involved subsistence commodities such as salt and food grains from Kashmir, wool (bal) and sheep from Changthang, and dried apricots (chuli) from Baltistan (Rizvi, 2004). The nomadic communities of Changthang and the adjoining highlands also took part in barter trade (tsonglen), exchanging Tibetan goods such as salt, medicine, gold, and shawl-wool for Indian items like cloth, sweets, rice, and barley (Warikoo, 1990). The pashm trade formed the political and commercial link of Ladakh with Kashmir (khachul) for centuries (Rizvi, 2004). The Srinagar-Leh-Yarkand route was one of the most significant, challenging, and long-established trade routes between India and Central Asia (Warikoo, 1989). It was along this route that the Kashmir shawl industry received its supplies of pashm wool, imported into Ladakh from Tibet and Yarkand (Warikoo, 1989). Two large caravan sarais at Leh and Srinagar accommodated South and Central Asian traders (Warikoo, 1989). Along the route, various facilities such as godowns, dak bungalows, stables, and inns were set up to support traders, offering them shelter as well as subsidised grain and fodder for their ponies (Warikoo, 1989). In addition, in April 1873, a treaty was concluded between Maharaja Ranbir Singh and T. Forsyth, which included a provision for the annual allocation of 2,500 rupees

[4] The family of Munshi Habibullah, one of Aziz Bhat's elder sons, established a museum to exhibit the diverse trade artefacts that passed through Kargil in the early twentieth century (Fewkes, 2008). This family-operated museum, known as the Munshi Aziz Bhat Museum, is one of the earliest regionally curated institutions that showcases Ladakh's contribution and active participation in the trade between South and Central Asia (Fewkes, 2008).

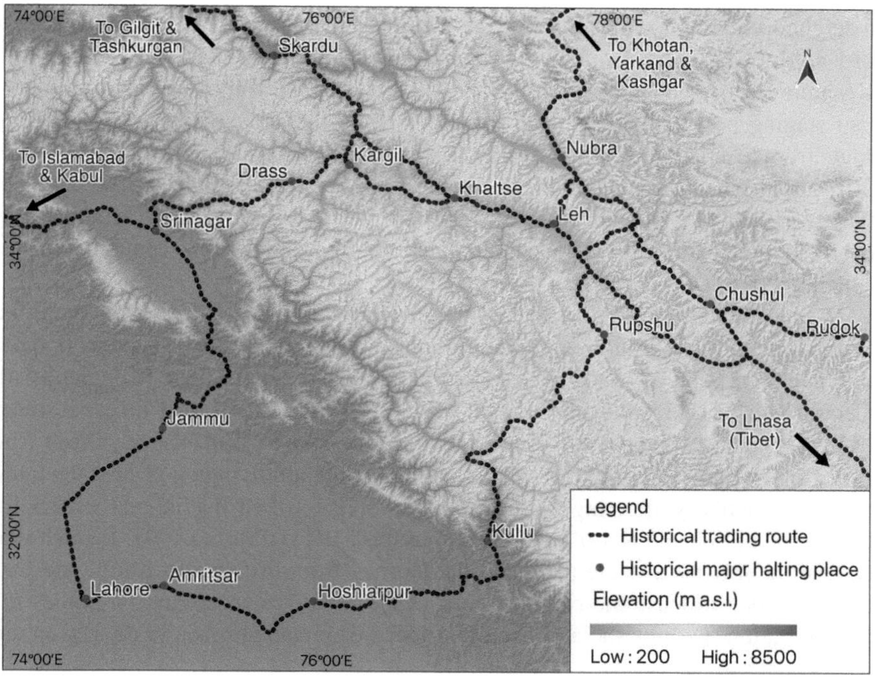

Fig. 4.3 Historical trading routes of the nineteenth and early twentieth centuries through Ladakh. *Sources* SRTM (11 February 2000); Fewkes (2008), Rizvi (2004). *Draft and cartography* Altaf Hussain

for the maintenance of roads and sarais along the Indo-Central Asian trade route, which was later widely recognised as the 'Treaty Road' (Warikoo, 1989).

All the commodities were carried on the backs of animals like horses, donkeys, mules, yaks, dzos, sheep and occasionally goats, and camels for transporting the trading goods (Joldan, 2006; Rizvi, 2004). The transporters (kiraiyakash) used to play a vital role in facilitating the shipment of goods between South Asia and Central Asia (Bray, 2005; Fewkes, 2008; Rizvi, 2004). The transporters (kiraiyakash[5]) were local villagers who possessed a deep understanding of regional conditions and were familiar with the terrain when transporting goods through Ladakh (Fewkes, 2008). Kargil was an important centre for the local trade (tsonglen) towards Skardu, the Suru Valley, as well as Leh and Srinagar (Rizvi, 2004). Local communities from surrounding areas in Kargil contributed labour and livestock, such as donkeys, ponies, and dzos, to transport goods to Leh (Fewkes, 2008). Kargil had a serai that included a trade (tsonglen) depot and an inn, where goods from all four directions were exchanged and stored (Rizvi, 2004). The Baltis in Kargil town are descendants of traders from the Kharmang region of Baltistan, and this trading community formed

[5] The kiraiyakash served as transporters of goods or providers of livestock for hire, and the majority of them belonged to rural communities in and around Kargil and Drass (Bray, 2005; Rizvi, 2004).

a neighbourhood known as Balti Bazaar, which is the oldest part of the main bazaar of Kargil town. Goods from all directions like Changthang, Zanskar, Baltistan, and Kashmir exchanged hands in Balti Bazaar (Gupta, 2022). Kargil was also an important staging post or halting place (brangsa) along the Treaty Road, a key trading route from Srinagar to Leh and further onward to Central Asia (Rizvi, 2012; Sheikh, 2014). It also had great economic and strategic importance because of its location at the crossroads of Baltistan, Kashmir, Zanskar, Leh and Central Asia (Mato Bouzas et al., 2019; Sheikh, 2014). Thus, it remained an important regional centre in the past (Sheikh, 2014) (Fig. 4.4).

When Ladakh served as a crucial trade route with Tibet and Central Asia, the inhabitants of the Kargil area, including those of Kharbu, often worked as porters, utilising their ponies to carry loads. However, the cessation of trade (tsonglen) with Central Asia in 1930 and with Tibet following its occupation by the Chinese resulted in significant disruption to the economy of the pony-owners (Zutshi et al., 1961). The cessation of trade had a serious impact on the population living on or near the trade routes, particularly affecting those who relied on it for their livelihoods, especially the people of Leh and Kargil in Ladakh (Wreford, 1943). Prior to 1947, Ladakh was accessible only by a bridle path, and the journey from Srinagar to Kargil and Leh typically took at least one or two weeks, respectively, often relying on ponies for transportation. Despite the absence of surfaced or metalled roads in the region, an

Fig. 4.4 Porters receiving expedition equipment in Leh, Ladakh. *Sources* Helmut De Terra, photograph taken on 16 June 1932, American Geographical Society Library, University of Wisconsin-Milwaukee Libraries

extensive programme to construct jeepable roads was initiated simultaneously. The Srinagar-Kargil-Leh road was completed and opened to light vehicular traffic on 1 August 1960 (Zutshi et al., 1961).

4.5 Emergence of Islam

Mir Syed Ali Hamadani, known as Shah-e-Hamadan, was the first to introduce Islam to Ladakh, converting locals during his journey to Kashgar in 1394 A.D. According to local tradition, he constructed mosques at Padum in Zanskar and Shey in Leh (Sheikh, 1989). Sayyid Muhammad Nurbakhsh, one of his disciples, brought Islam to Purig (Kargil) as well as to Baltistan, and his followers are known as Noorbakshi in both Ladakh and Baltistan (Bray, 1991; Sheikh, 1989). Syed Nurbaksh and his son Shah Qasim Faiz Baksh were instrumental in the extensive propagation of Islam beyond the Zoji La in Ladakh (Khan, 1912). The conversion of the people of Baltistan and Kargil from Tibetan Buddhism to Islam began in the fifteenth century (Gupta, 2014). Most parts of these regions initially adopted Nurbakhshi Islam in the early fifteenth century under the influence of Syed Mohammad Nurbakhsh's followers (Gupta, 2014). In 1505, Shamsud Din Iraqi, a prominent religious scholar, converted many people to Shia Islam in Purig (Kargil) as well as in Baltistan (Sheikh, 1989). The Ladakh kingdom experienced repeated invasions by the Kashmir Sultanate beginning in 1405, followed by a Mughal incursion led by Mirza Haidar Dughlat, whose forces occupied the region between 1532 and 1536 (Bray, 2012; Dollfus, 1995). More important than the dissemination of Islam, commercial interests attracted Mirza Haidar Dughlat to the Tibetan plateau. In 1548, he seized Baltistan and Ladakh to ensure the security of the trade routes, and to develop trade in Central Asia (Dollfus, 1995). In the late sixteenth century, Ladakh's King Jamyang Namgyal (1595–1616) was captured following a military invasion. He is traditionally believed to have fallen in love with Gyal Khatun, the daughter of Ali Mir (Ali Sher Khan Anchan) of Skardu, a renowned Balti queen of Baltistan (Bray, 2012; Halkias, 2011). In the seventeenth and eighteenth centuries, the kingdoms of Purig and Ladakh were frequently engaged in battles over control of trade with Kashmir. During these confrontations, the Purig chiefs often aligned with the chiefs of Baltistan, with whom they shared a common religious background (Petech, 1939, 1977). Following the 1947 partition of India and Pakistan, Ladakh became part of the Indian state of Jammu and Kashmir, whereas Baltistan was separated from India and integrated into Pakistan's Northern Areas.

4.6 Post-independence Period

Ladakh remained an independent kingdom from about 950 A.D. until 1834, when it was invaded by the Dogras and came under the control of the Maharaja of Jammu and Kashmir. During the nineteenth and early twentieth centuries, it served as a vital crossroads for Trans-Himalayan and Trans-Karakoram trade, linking Tibet, India, Afghanistan, and Central Asia (Fewkes, 2008; Rizvi, 2012). However, the situation of cross-border trade among these regions changed significantly after the partition of India and the end of British rule. The disruption of Ladakh's cultural, religious, and economic connections with Tibet, Baltistan, and Central Asia, combined with its integration into India, introduced new forms of modernity and development but also resulted in marginalisation and discrimination (Beek, 2000). After the 1947 partition of India and Pakistan, Ladakh was integrated into the Indian state of Jammu and Kashmir. The road connectivity between Kargil and Skardu was closed after the partition and made Ladakh entirely dependent on the Kashmir Valley for basic requirement of goods, education, health care, and politics.

Ladakh was accessible only by a bridle path, and the journey from Srinagar to Kargil and Leh was made on ponies (Zutshi et al., 1961). There was no road connectivity in Ladakh. When the Srinagar-Kargil-Leh road was first opened to motorised vehicles in 1960, it began to bring significant socio-economic transformation (Demenge, 2011; Zutshi et al., 1961). The Sino-Indian War of 1962 and territorial conflicts with Pakistan led to the development of the Srinagar-Kargil-Leh highway (NH-1) and the Manali-Leh highway (NH-3) between the 1960s and 1970s, primarily for military purposes, later opened for civilian transport (Baghel & Nüsser, 2015; Demenge, 2011; Kreutzmann, 1991; Zutshi et al., 1961). The building of these major roads acted as a catalyst for the transformation of Leh and Kargil in Ladakh. In addition, the Kargil district of Ladakh was the site of hostilities during the armed conflicts between India and Pakistan in 1947, 1965, and 1971. During the 1971 war, the Indian Army gained control of the village of Turtuk in the Shyok Valley, which had previously been controlled by Pakistan (Królikowski, 2021). The region was also a major theatre of conflict during the Sino-Indian War of 1962, which resulted in China claiming the Indian-held Aksai Chin region (Vogel & Field, 2020). The 1965 and 1971 conflicts between India and Pakistan brought Kargil district to the forefront of hostilities, primarily due to unresolved issues over Kashmir and the Siachen glacier (Adnan, 2015; Baghel & Nüsser, 2015; Vogel & Field, 2020). National security concerns played a major role in shaping the functioning of the civil administration in Ladakh (Aggarwal & Bhan, 2009). In the aftermath of the Sino-Indian War, a development committee was formed in 1965 to address key areas such as education, electricity, and irrigation, and to mitigate regional disparities within the state (Aggarwal & Bhan, 2009). In the subsequent years, as civil insurgency escalated in Kashmir, Ladakh was granted Scheduled Tribe status in 1989, followed by the establishment of the LAHDC in 1995 (Aggarwal & Bhan, 2009).

The prolonged political conflict between India and Pakistan escalated into an undeclared war in Kargil in 1999, resulting in 474 deaths (Kapur, 2008; Zins, 2007). Moreover, nearly 35,000 people were displaced or fled from their homes in Kargil, Pandras, Matayan, Drass, and Batalik due to Pakistan shelling (Warikoo, 1999; Zins, 2007). Kargil Vijay Diwas (Kargil Victory Day) is observed annually on 26 July to commemorate India's victory over Pakistan in the 1999 Kargil War and to honour the soldiers who fought for the nation. The military presence in Ladakh, which intensified in the early 1960s, opened new avenues of employment for Ladakhis (Wiley,

1998). After the opening to tourism in 1974, the income and local economy of the people in both Leh and Kargil flourished in Ladakh. A significant shift took place from agriculture-based livelihoods to monetary incomes, primarily driven by the growth of the tourism sector (Dame et al., 2019; Geneletti & Dawa, 2009; Goeury, 2010). The tourism sector, road connectivity, government, and army activities have brought massive political and economic changes in terms of employment opportunities, economic development, and political representation in Kargil (Bhan, 2009, 2013; Grist, 2008).

4.7 Administration and Governance

Ladakh has experienced a significant administrative transformation in its governance system. In 1979, Kargil was separated from the erstwhile Leh district and became an independent district of Ladakh during the tenure of the state's Chief Minister (Sheikh Abdullah). The people of Kargil argued that the separation was essential to enhancing administrative efficiency and ensuring a fair allocation of resources between Leh and Kargil. A vast majority of Shia Muslims in Kargil perceived themselves as isolated and underdeveloped due to the state government's indifferent and neglectful attitude towards them (Bhan, 2013). However, the division sparked tensions regarding resource allocation, economic progress, and Ladakh's political relationship with Kashmir (Bhan, 2009). The delineation of the Kargil district broadly followed the administrative division between Leh and Kargil tehsils, and Kargil became a Muslim-majority district with Zanskar a Buddhist minority region within it (Gupta, 2022). In response to the agitation that began in 1989, the LAHDC Act was enacted in 1995, granting administrative authority over decision-making and financial matters, while excluding jurisdiction over law and order and the judiciary (Aggarwal, 2004). A semi-autonomous body, the Ladakh Autonomous Hill Development Council, was established in 1993. It was introduced in Leh in 1995 (LAHDC-Leh) and subsequently in Kargil in 2003 (LAHDC-Kargil) (Beek, 2000; Bhan, 2013). When the first LAHDC-Leh was established on 3 September 1995, Leh district, then part of the state of Jammu and Kashmir, regained a measure of the autonomy that the Ladakh kingdom had lost in the 1830s (Beek, 1998). The council was implemented only in the Buddhist-majority Leh district, while the Muslim-majority Kargil rejected the offer. The LAHDC was entangled in religious and political conflicts from its inception in 1989 until its implementation in 1995 (Bhan, 2009). The political rift between Kargil and Leh widened in 1989 when the Ladakh Buddhist Association (LBA) initiated a mass movement demanding UT status, advocating for Ladakh's separation from the Kashmir region (Bhan, 2013). The Buddhist leadership in Leh accused the Jammu and Kashmir government of implementing biased policies towards Ladakh's cultural and religious minorities (Bhan, 2013). In 2003, under the leadership of Mufti Mohammad Sayeed (People's Democratic Party), the Jammu and Kashmir government, in line with its 'healing touch' policy, initiated elections for the LAHDC-Kargil. This initiative aimed to promote governance, participatory development, and foster democratic practices in the conflict-affected district (Bhan, 2013). The LAHDC-Kargil holds administrative authority over governance and district development planning, advancing nation-building efforts in the borderland frontier through mechanisms of decentralisation, participatory development, and self-governance (Bhan, 2013) (Fig. 4.5).

Ladakh is now governed by the UT administration under the Lieutenant Governor, who functions as a representative of the central government of India. Unlike other Union Territories with legislative assemblies, Ladakh lacks an elected assembly but retains a seat in the Lok Sabha, the lower house of Parliament. During the erstwhile state of Jammu and Kashmir, Ladakh had legislative representation, with the right to elect four members to the state assembly, nominate two members to the legislative council, and appoint a secretary to manage its regional affairs. Besides the LAHDCs, the two districts (Leh and Kargil) have district development commissioners who are responsible for implementing government schemes and supervising daily affairs in Ladakh. According to the LAHDC Act of 1995, the semi-autonomous body comprises 30 members, of whom 26 are elected, while the remaining 4 seats are reserved for women and nominated representatives from the district's principal religious minority.

Ladakh was granted divisional status within the state of Jammu and Kashmir in February 2019. Subsequently, in October 2019, the former state was reorganised into two distinct Union Territories: Ladakh and Jammu and Kashmir. Ladakhi Buddhists celebrated the UT declaration, while the Muslims of Kargil vociferously and unanimously rejected it. Soon after the announcement, Kargil political parties, along with Islamia School Kargil (ISK) and Imam Khomeini Memorial Trust (IKMT), continued to mediate politics in the district. These groups came together to form the

Fig. 4.5 Newly established secretariat of LAHDC-Kargil at Kurbathang Plateau, Kargil, Ladakh. Photograph taken by Altaf Hussain, 11 September 2021

Kargil Joint Action Committee, which called for a two-day strike and demanded the restoration of Articles 370 and 35A, as well as the lifting of the curfew and communication blockade (Gupta, 2022). Hundreds of people across the district (excluding the Buddhist-majority Zanskar) participated in the protests, defied the imposition of Section 144 of the Indian Penal Code, which prohibits large gatherings, denounced the actions of the Government of India, and clearly expressed their allegiance to Kashmir (Gupta, 2022). Kargili leadership across the political and religious spectrum (excluding the local BJP party) came together to establish Kargil Democratic Alliance (KDA) in October 2020. The KDA demanded the restoration of Article 370 and called for statehood for Ladakh to restore democratic rights and representation. When the leadership in Leh received little response from the union government regarding their demands for 6th Schedule status, they also began advocating for statehood by forming Leh Apex Body (LAB) in Ladakh. Both KDA and LAB united on a common platform and worked together for the political and democratic rights of Ladakh. LAB and KDA are two alliances of representatives from socio-religious and political organisations, striving to fulfil the political demands of the people of Ladakh through negotiations with central government of India. The full restoration of erstwhile statehood, political representation in the Lok Sabha and Rajya Sabha, reservation of jobs for Ladakhis, and constitutional safeguards under 6th Schedule are the major political demands of the people of Ladakh.

Furthermore, after negotiations and the central government's failure to address key demands, hundreds of demonstrators gathered in Kargil and Leh in March 2024 to demand the fulfilment of Ladakh's four main concerns. Led by the KDA, prominent political and religious leaders, including Asgar Ali Karbalai, Qambar Ali Akhoon, Sheikh Hassanain Rizwani, and Sajjad Hussain Kargili, addressed the crowd at Hussaini Park and expressed support for the movement against the central government. Asgar Ali Karbalai announced a hunger strike from 24 to 27 March, organised by the KDA, and urged locals to join the protests. The demonstrators sought constitutional protections to safeguard the region's fragile ecology and autonomy over land and agricultural matters. Additionally, the political and socio-religious leaders organised protests in Leh to advocate for Ladakh's major political demands. Sonam Wangchuk, a climate activist, ended his 21 day hunger strike in Leh, which he undertook to urge the central government to safeguard the region's fragile environment and to demand its inclusion under the 6th Schedule of the constitution. Thus, the administrative landscape in Ladakh has experienced significant political shifts due to changing geopolitical dynamics and structural governance reforms.

4.8 Contemporary Socio-religious Development

4.8.1 Religious Development

The district of Kargil has two main socio-religious organisations, namely ISK and IKMT. These institutions hold a prominent position in the religious, social, and political life of the people of Kargil, Ladakh. ISK, also known as Jamiat-ul-Ulama Isna Asharia Kargil, is the oldest and largest religious organisation in the region. It was established in 1953 by Sheikh Mohammad Mufeed of Chutumail, with generous support from the local community (Milton, 2013). The institution grew into a highly respected establishment under the leadership of Sheikh Ahmed Mohammadi, who was instrumental in shaping ISK into a prominent and influential religious organisation. Over the years, ISK has played a pivotal role in promoting religious education, cultural preservation, and community development in the Kargil, Ladakh.

IKMT is the offshoot of ISK, founded on July 13, 1989, in Kargil under the leadership of Sheikh Mohd Hussain Zakiri and Asgar Ali Karbalai (former Chief Executive Councillor of LAHDC-Kargil). Its main objectives were to strengthen the Islamic spirit, promote women's empowerment, and spread modern education among the youth, particularly in the far-flung areas of the district (Gupta, 2022; Milton, 2013). The founders of the IKMT dissociated themselves from the principal religious institution of the ISK, arguing that its conservative orientation was impeding the progress of Kargil society (Gupta, 2014). With the founding of the IKMT, Kargili society became deeply divided between the politico-religious factions (Gupta, 2022). In 1997, tensions arose between ISK and IKMT over religious practices during the Ashura procession. IKMT questioned the tradition of self-flagellation with chains and blades (Zanjiri Matam), advocating instead for blood donation in line with the 1994 guidance of Ayatollah Syed Ali Khamenei, the Supreme Shia leader of Iran (Aggarwal, 2004; Milton, 2013). However, ISK supporters opposed these efforts, considering bloodletting rituals and mourning practices as deep expressions of their devotion to the Prophet's family and a means of expressing solidarity with their pain and sacrifice (Aggarwal, 2004).

IKMT is committed to upholding the legacy and teachings of Ayatollah Ruhollah Khomeini, the spiritual leader of the 1979 Iranian Revolution and the founding father of the Islamic Republic of Iran. It often engages in various educational, healthcare, philanthropic, and religious activities. Influenced by the Islamic revolution of 1979, IKMT decided to intervene in the socio-economic life of the local people by initiating a number of development programmes (Milton, 2013). In health care, the modern mindset of IKMT has led to the formation of Baqirya Health Centre and Research Centre. This health centre comprises doctors, educationists, engineers, and other professionals committed to providing free medical consultations, conducting medical camps in remote areas, organising medical awareness programmes, and assisting government agencies in implementing various health missions. To provide modern, quality education, more than 12 Mutahhary public schools have been established in various parts of Kargil, all of which are run by the Mutahhary Educational

Society of IKMT. Besides promoting Islamic teachings, it also fosters unity and solidarity among all social and religious communities in Ladakh. Additionally, ISK has initiated efforts in modern education, the medical field, and other social issues, with a particular focus on the less privileged members of the local communities in Kargil, Ladakh. In the 1970s, certain Shi'ite leaders in Kargil discouraged their followers from pursuing secular education and cautioned against the use of English and Urdu languages (Grist, 2008). To provide modern education in Kargil, the first private school, Suru Valley Public School, was set up in 1982 by Reverend Gergan from the Moravian Mission (Gupta, 2022). Additionally, Khomeini's revolutionary ideologies have brought significant changes to various aspects of social, religious, and political life, leading ISK to adopt a more modernist and progressive approach (Gupta, 2014). ISK has traditionally adhered to an Iraq-oriented ideological position, aligning with the views of Ayatollah Sistani. However, over the years, IKMT has emerged as a prominent alternative religious and political centre in Ladakh (Gupta, 2014).

4.8.2 Operation Sadbhavana

In the aftermath of the Kargil War, Lieutenant General Arjun Ray assumed command of the 14th Corps, a unit within the Indian Army's Northern Command established specifically for the Ladakh region. In 2001, he introduced Operation Sadbhavana (Operation Goodwill), an initiative designed to foster cooperation among border communities. The operation aimed to build trust and goodwill by promoting human development in these strategically borderland areas (Aggarwal & Bhan, 2009). Under this mission, the Indian army implemented projects to promote primary education, vocational training, community development, women's empowerment, horticulture, healthcare centres, and so on (Bhan, 2013; Suri, 2013). Operation Sadbhavana addressed development challenges in Ladakh through a framework grounded in grassroots empowerment, sustainability, and locally driven initiatives (Aggarwal & Bhan, 2009). This welfare and development initiative was funded through allocations from the central government's Border Area Development Fund, along with support from the Ministry of Defence. Its objectives included promoting peacebuilding, enhancing human security, and improving development infrastructure for border communities, while also instilling a sense of patriotism (Aggarwal & Bhan, 2009; Bhan, 2013). In addition, the Indian military benefited from financial and logistical assistance provided by other entities, including the Ministry of Information Technology and private IT firms such as IBM and Infosys (Bhan, 2013). Volunteers trained in information technology were invited to participate in Sadbhavana initiatives from India's Silicon Valley in Bangalore. As a war zone site, Kargil became deeply embedded in the national consciousness, prompting calls for youth from metropolitan areas to contribute their skills in service of the nation (Aggarwal & Bhan, 2009).

To address educational needs, the Indian military initiated the establishment of primary education centres across Ladakh. By the end of 2002, it had set up 16 Sadbhavana schools aligned with the Central Board of Secondary Education curriculum (Bhan, 2013). In the healthcare sector, Operation Sadbhavana sought to deliver primary, secondary, and tertiary medical services to the Ladakhi population through a network of medical aid posts, mobile field ambulances, and hospitals located in Leh, Delhi, and Chandigarh (Bhan, 2013). The initiative also prioritised women's empowerment by setting up training centres that offered instruction in knitting, weaving, and sewing. Furthermore, programmes promoting adult education and computer literacy were introduced within these centres in Ladakh (Bhan, 2013).

The 1999 Kargil War highlighted the challenges of warfare in high-altitude terrain and underscored the crucial role of local soldiers, which ultimately led to the Ladakh Scouts being granted full regimental status. Initially a small unit with only a few battalions during the 1962 Sino-Indian War, the Ladakh Scouts were reorganised in 2000 by the Indian government, with new regiments established to boost local recruitment across Ladakh (Aggarwal & Bhan, 2009; Bhan, 2008). Following the Kargil conflict, the presence of Indian troops in the region surged from around 3,000 to nearly 20,000, further embedding the military in the everyday lives and economic activities of Ladakhi communities (Aggarwal & Bhan, 2009). For Brogpas, military recruitment became the most accessible occupation particularly in the border areas of Garkone, Dartsigs, Batalik, Lalung, and other neighbouring villages of Kargil (Bhan, 2008).

The setting up of a military base in the Batalik sector created new income-generating opportunities for the Brogpa community, who began working as salaried porters. The support they provided to the army during times of crisis (khral or mandatory labour) gradually turned into a primary occupation for them as the conflict over Kashmir between India and Pakistan escalated (Bhan, 2008). The growing and diverse presence of the Indian Army in Ladakh enabled many locals to secure employment in paramilitary units like the Ladakh Scouts and the Indo-Tibetan Border Force, as well as in support roles such as porters and cooks (Aggarwal & Bhan, 2009). Since the Sino-Indian War of 1962, the relationship between civilians and the Indian military has been an integral component of Ladakh's political economy.

References

Adair, F. E. S., & Godfrey, S. H. (1899). *A summer in High Asia: Being a record of sport and travel in Baltistan and Ladakh, with an appendix on Central Asian trade, illustrated from drawings by the author, photographs, and a map of the route.* W. Thacker & Company.

Adnan, M. (2015). The Kargil crisis 1999 and Pakistan's constraints. *Journal of Political Studies, 22*(1), 129.

Aggarwal, R. (2004). *Beyond lines of control performance and politics on the disputed borders of Ladakh, India.* Duke University Press.

Aggarwal, R., & Bhan, M. (2009). "Disarming Violence": Development, democracy, and security on the borders of India. *The Journal of Asian Studies, 68*(02), 519. https://doi.org/10.1017/S0021911809000692

Aitchison, J. E. T. (1874). *Hand-book of the trade products of Leh, with the statistics of the trade, from 1867 to 1872 inclusive*. Wyman & Co.

Baghel, R., & Nüsser, M. (2015). Securing the heights: The vertical dimension of the Siachen conflict between India and Pakistan in the Eastern Karakoram. *Political Geography, 48*, 24–36. https://doi.org/10.1016/j.polgeo.2015.05.001

Bhan, M. (2008). Border practices: Labour and nationalism among Brogpas of Ladakh. *Contemporary South Asia, 16*(2), 139–157. https://doi.org/10.1080/09584930701733472

Bhan, M. (2009). Refiguring rights, redefining culture: Hill-councils in Kargil, Jammu and Kashmir. *Sociological Bulletin, 58*(1), 71–93. https://doi.org/10.1177/0038022920090105

Bhan, M. (2013). *Counterinsurgency, democracy, and the politics of identity in India*. Routledge. https://doi.org/10.4324/9781315889863

Bray, J. (1983). Moravian church in Ladakh: The first forty years 1885–1925. In D. Kantowsky & R. Sander (Eds.), *Recent research on Ladakh* (pp. 81–91). Weltforum Verlag.

Bray, J. (1991). Ladakhi history and Indian nationhood. *South Asia Research, 11*(2), 115–133. https://doi.org/10.1177/026272809101100201

Bray, J. (Ed.). (2005). *Ladakhi histories: Local and regional perspectives*. Brill.

Bray, J. (2008). August Hermann Francke's letters from Ladakh 1896–1906: The making of a missionary scholar. *Library of Tibetan Works and Archives, 33*(1), 3–28. https://www.jstor.org/stable/43301170

Bray, J. (2012). Readings on Islam in Ladakh: Local, regional, and international perspectives. *Himalaya, the Journal of the Association for Nepal and Himalayan Studies, 32*(1), 10. http://digitalcommons.macalester.edu/himalaya/vol32/iss1/9

Chand, R. (2000). Agricultural development, growth and poverty in India's mountain region. In M. Banskota, T. S. Papola, & J. Richter (Eds.), *Proceedings of the international conference held from 31 January–4 February 2000 in Kathmandu, Nepal* (pp. 275–291). International Centre for Integrated Mountain Development and German Foundation for International Development.

Cunningham, A. (1854). *Ladák, physical, statistical, and historical: With notes of the surrounding countries*. Wm. H. Allen and Company.

Dame, J., Schmidt, S., Müller, J., & Nüsser, M. (2019). Urbanisation and socio-ecological challenges in high mountain towns: Insights from Leh (Ladakh), India. *Landscape and Urban Planning, 189*, 189–199. https://doi.org/10.1016/j.landurbplan.2019.04.017

Dani, A. H. (1991). *History of northern areas of Pakistan* (2nd ed.). National Institute of Historical and Cultural Research.

Datta, C. L. (1973). *Ladakh and Western Himalayan politics, 1819–1848: The Dogra conquest of Ladakh, Baltistan, and West Tibet, and reactions of the other powers*. Munshiram Manoharlal Publishers.

De Filippi, F. (1912). *Karakoram and Western Himalaya 1909: An account of the expedition of HRH Prince Luigi Amedeo of Savoy, Duke of the Abruzzi*. E. P. Dutton and Company.

Demenge, J. (2011). *The political ecology of road construction in Ladakh*. University of Sussex. http://sro.sussex.ac.uk/38501/

Dollfus, P. (1995). The history of Muslims in Central Ladakh. *Library of Tibetan Works and Archives, 20*(3), 35–58. https://www.jstor.org/stable/43300542

Fewkes, J. H. (2006). Trade at the crossroads of continents: Northern India's Ladakh region. *Expedition (Penn Museum), 48*(3), 18–26. https://www.penn.museum/sites/expedition/trade-at-the-crossroads-of-continents/

Fewkes, J. H. (2008). *Trade and contemporary society along the silk road*. Routledge. https://doi.org/10.4324/9780203273692

Francke, A. H. (1926). F. W. Thomas (Ed.), *Antiquities of Indian Tibet: Part II-the chronicles of Ladakh and minor chronicles, texts and translations, with notes and maps* (Vol. L). Superintendent Government Printing, India.

Francke, A. H. (1995). *A history of Western Tibet: One of the unknown empires with maps and illustrations.* Asian Educational Services.

Gazetteer of Kashmir and Ladák. (1890). *Together with routes in the territories of the Maharája of Jamú and Kashmír; compiled under the direction of the Quarter Master General in India in the Intelligence Branch.* Superintendent of Government Printing, India.

Geneletti, D., & Dawa, D. (2009). Environmental impact assessment of mountain tourism in developing regions: A study in Ladakh, Indian Himalaya. *Environmental Impact Assessment Review, 29*(4), 229–242. https://doi.org/10.1016/j.eiar.2009.01.003

Goeury, D. (2010). Ladakh, kingdom of sustainable development?: Protecting the natural environment to protect identity. *Revue de Géographie Alpine, 98-1.* https://doi.org/10.4000/rga.1147

Grist, N. (2008). Urbanisation in Kargil and its effects in the Suru Valley. In M. van Beek, & F. Pirie (Eds.), *Modern Ladakh: Anthropological perspectives on continuity and change.* Brill. https://doi.org/10.1163/ej.9789004167131.i-313

Gupta, R. (2014). Experiments with Khomeini's revolution in Kargil: Contemporary Shi'a networks between India and West Asia. *Modern Asian Studies, 48*(2), 370–398. https://doi.org/10.1017/S0026749X13000759

Gupta, R. (2022). *Freedom in captivity: Negotiations of belonging along Kashmir's frontier.* Cambridge University Press.

Halkias, G. T. (2011). The Muslim Queens of the Himalayas: Princess exchanges in Baltistan and Ladakh. In A. Akasoy, C. Burnett, & R. Yoeli-Tlalim (Eds.), *Islam and Tibet-interactions along the musk routes.* Ashgate.

Huttenback, R. A. (1973). The emasculation of a princely state: the case of Kashmir. *Journal of Asian History, 7*(1), 1–29. https://www.jstor.org/stable/41930070

Joldan, S. (2006). Relationship between Ladakh and Buddhist Tibet: Pilgrimage and trade. *Library of Tibetan Works and Archives, 31*(3), 43–76. http://www.jstor.org/stable/43300982

Kapur, S. P. (2008). Ten years of instability in a nuclear South Asia. *International Security, 33*(2), 71–94.

Khan, M.-Z. (1912). *Census of India, 1911: Kashmir, Part I-Report, Vol. XX.* Newul Kishore Press

Kreutzmann, H. (1991). The Karakoram highway: The impact of road construction on mountain societies. *Modern Asian Studies, 25*(4), 711–736. https://doi.org/10.1017/S0026749X00010817

Kreutzmann, H. (2008). Kashmir and the northern areas of Pakistan: Boundary-making along contested frontiers. *Erdkunde*, 201–219.

Królikowski, H. (2021). Indo-Pakistani "hybrid war" for Kargil. *Politeja, 13*(1(40)), 395–414. https://doi.org/10.12797/Politeja.13.2016.40.24

Mato Bouzas, A., Schendel, W., & Harris, T. (2019). *Kashmir as a borderland: The politics of space and belonging across the line of control.* Amsterdam University Press. https://doi.org/10.5117/9789463729406

Milton, N. (2013). *Kargil: An abode of mountain Tamers.* Har-Anand Publications Pvt Ltd.

Khan, K. B. M. G. A. (1902). *Census of India, 1901: Kashmir, Part I-Report, Vol. XXIII.* The Civil & Military Gazette Press.

Petech, L. (1939). *A study on the chronicles of Ladakh (Indian Tibet).* Calcutta Oriental Press.

Petech, L. (1977). *The Kingdom of Ladakh C. 950–1842 A.D.* Istituto Italiano per il Medio ed Estremo Oriente.

Rizvi, J. (1985). Peasant-traders of Ladakh: A study in oral history. *India International Centre Quarterly, 12*(1), 13–27. http://www.jstor.org/stable/23001353

Rizvi, J. (2004). *Trans-Himalayan caravans: Merchant Princes and peasant traders in Ladakh.* Oxford University Press.

Rizvi, J. (2012). *Ladakh: Crossroads of High Asia* (3rd ed). Oxford University Press.

Schettler, M., & Schettler, R. (1989). *Kashmir, Ladakh & Zanskar: A travel survival kit* (3rd ed.). Lonely Planet.

Sheikh, A. G. (1989). A brief history of Muslims in Ladakh. In H. Osmaston, & P. Denwood (Eds.), *Proceedings of the fourth and fifth international colloquia on Ladakh.* Motilal Banarsidass.

Sheikh, A. G. (2014). *Reflections on Ladakh, Tibet and Central Asia.* Inamullah Abdulmatin.

Suri, K. (2013). Women empowerment, conflict transformation and social change in Kargil. *International Journal of Social Science, 2*(2), 119–127. https://doi.org/10.5958/j.2321-5771.2.2.005

van Beek, M. (1998). True patriots: Justifying autonomy for Ladakh. *Himalaya, the Journal of the Association for Nepal and Himalayan Studies, 18*(1), 9. https://digitalcommons.macalester.edu/himalaya/vol18/iss1/9

van Beek, M. (2000). Beyond identity fetishism: "Communal" conflict in Ladakh and the limits of autonomy. *Cultural Anthropology, 15*(4), 525–569.

Vogel, B., & Field, J. (2020). (Re)constructing borders through the governance of tourism and trade in Ladakh, India. *Political Geography, 82*, 102226. https://doi.org/10.1016/j.polgeo.2020.102226

Warikoo, K. (1989). Gateway to Central Asia: The Trans-Himalayan trade of Ladakh, 1846–1947. In H. Osmaston, & P. Denwood (Eds.), *Proceedings of the fourth and fifth international colloquia on Ladakh.* Motilal Banarsidass.

Warikoo, K. (1990). Ladakh's trade relations with Tibet under the Dogras. *China Report, 26*(2), 133–144. https://doi.org/10.1177/000944559002600202

Warikoo, K. (1999). Kargil conflict view from Kashmir. *Himalayan and Central Asian Studies, 3*(3/4), 28–45.

Warikoo, K. (2014). The making of a frontier: The relationship between Kashmir and its frontier territories. In K. Warikoo (Ed.), *The other Kashmir: Society, culture and politics in the Karakoram Himalayas*. Pentagon Press.

Wiley, A. S. (1998). The ecology of low natural fertility in Ladakh. *Journal of Biosocial Science, 30*(4), 457–480. https://doi.org/10.1017/S002193209800457X

Wreford, R. G. (1943). *Census of India, 1941. Jammu and Kashmir: Essay and Tables, Part I and II, Vol. XXII*. Census Commissioner, Jammu & Kashmir State. https://dspace.gipe.ac.in/xmlui/handle/10973/37337

Zins, M.-J. (2007). Public rites and patriotic funerals: The Heroes and the Martyrs of the 1999 Indo-Pakistan Kargil War. *India Review, 6*(1), 25–45. https://doi.org/10.1080/14736480601172675

Zutshi, J. N., Farooqi, B. A., Din, G. M., & Nanda, J. K. (1961). *Census of India, 1961: Village survey monograph of Kharboo (Tehsil Kargil, District Ladakh), Part VI, No. 13, Vol. VI*. Director of Census Operations, Jammu & Kashmir.

Chapter 5
Urban Landscape Dynamics

Abstract This chapter offers an in-depth analysis of the urban landscape transformation of Kargil town over time. It explores several key aspects, including the historical evolution of Kargil, significant demographic changes, patterns of urban expansion, and the burgeoning tourism sector. The analysis considers the challenges and opportunities posed by the town's mountainous geography, which has a significant impact on its urban development. The chapter provides valuable insights into the processes driving urbanisation, highlighting how the mountainous terrain shapes infrastructure development, building patterns, and the overall urban landscape structure. Furthermore, it delves into the socio-economic factors influencing these changes, such as rural-to-urban migration, economic opportunities, and the rise of tourism as a major economic driver. By examining these multifaceted elements, the chapter offers a detailed understanding of Kargil's urban landscape dynamics, emphasising the interplay between natural and human factors in shaping the town's development.

Keywords Urbanisation · Urban population · Urban expansion · Built-up area · Settlements · Rural-to-urban migration · Unplanned development · Tourism sector · Tourists · Kargil town · Ladakh

5.1 Evolution of Kargil Town

In the ancient times, Purig consisted of several small kingdoms like Kartse, Mulbekh, Sodh, Pashkyum, Fokar, Shimsha Kharbu, and Chiktan (Census of India, 2011b; Cunningham, 1854; Francke, 1926; Khan, 1992; Petech, 1977). During the ninth century, Tha Tha Khan, a scion of the Gilgit royal family, migrated to Purig, established his kingdom, and brought the surrounding small kingdoms, including Kartse,

A. Hussain, *Urbanisation and Development Challenges in Kargil, Ladakh*, Advances in Asian Human-Environmental Research, https://doi.org/10.1007/978-3-032-05667-2_5

Pashkyum, Chiktan,[1] and Sodh, under his rule (Census of India, 2011a; Vohra, 1989). The chieftains of Sodh were the descendants from Tha Tha Khan (Khra Khra Khan), and he was in possession of three castles (Francke, 1926). The Pasar and Sodh castles were situated at Yourbaltak (Sodh), with the Pasar castle (khar) also known as Mkhah-hdror or the fairy castle (Francke, 1926).

Prior to the nineteenth century, Kargil served as the capital of the Purig district (Bellew, 1875; Duke, 1903). It was a substantial village (yul), situated on densely wooded slopes overlooking the river, with the fertile agricultural lands of Poyen visible across the opposite bank (Bellew, 1875; Gazetteer of Kashmir and Ladák, 1890). There were scattered villages on either side of the Suru River (Neve, 1918). Wheat (kro), barley (nas), apricots (chuli), mulberries (owse), willows (malchang), and poplars (zberpa) flourished along the watercourses of the terraced fields (Drew, 1875; Duke, 1903; Duncan, 1906). The region was inhabited by Ladakhis who had converted to Islam and were under the jurisdiction of the Baltistan administration based in Skardu (Neve, 1918).

It functioned as the chief place (Schlagintweit et al., 1861) and had a fort, collectorate, large serai, rest-house, polo ground, post office, and telegraph office (Bellew, 1875; Duke, 1903; Gazetteer of Kashmir and Ladák, 1890; Neve, 1918). The collectorate was situated at the top of the settlement, which comprised some neat and commodious buildings (Bellew, 1875; Gazetteer of Kashmir and Ladák, 1890). According to Filippi, "On the hillside, are the serai, the post and telegraph office, a fine large bungalow, and a fine small house belonging to the Wazir-i-Wazarat. A little farther on, a tributary flows into the Suru from the east, and a big Dogra fortress stands nearby on the left bank" (Filippi, 1931, p. 105). Furthermore, Kargil was considered a wonderful place, characterised by a single hilly street lined with small wooden shops and large emporiums. These establishments offered goods such as matches, kerosene oil, various types of sugar (khara) and tea (cha), cotton fabrics from Bombay and Manchester, along with inexpensive glass and tinsel ornaments (Gompertz, 1926) (Fig. 5.1).

The main settlements near Kargil included Mulbekh, Pashkyum, and Sodh, and each of which once had its own petty chief (Cunningham, 1854). Pashkyum had a handful of shops that sold basic items such as flour (phey), rice (bras), butter, and other provisions (Moorcroft & Trebeck, 1841), and consisted of a number of houses scattered along two valleys, disposed in terraces, and well cultivated with lucerne (ol), wheat (kro), and barley (nas) (Moorcroft & Trebeck, 1841). Sodh consisted solely of a fort, located on the right bank of the Suru River, to the east of Kargil (Moorcroft & Trebeck, 1841). In June 1875, a sub-post office was opened in Leh, followed by one in Kargil. Additionally, 32 dak stages were established from Sonamarg to Leh, with two men appointed to each stage during the summer, while 16 men were employed between Sonamarg and Matayen during the winter (Sheikh, 2014). In 1912, the total

[1] Until the early nineteenth century, Chiktan (Chigtan), located in the Purig region of Western Ladakh, was a semi-autonomous principality governed by a local chief. Like the rest of Purig, it was originally a Buddhist-dominant area, but the ruling family converted to Islam in the late sixteenth or early seventeenth century (Khan et al., 2014).

Fig. 5.1 Views of the old and existing bazaars along the same street in Kargil town. This street, which was once the main bazaar for trading activities, has now transformed from a bustling marketplace into an inactive, deserted area. *Sources* old photo from May 1931 (Bates & Harman, 2014); new photo taken by Altaf Hussain, 18 September 2021

number of government officials in Skardu, Leh, and Kargil was 165, according to the Leh archives, and the majority of the employees were from Jammu, Kashmir, and Punjab, rather than Ladakh (Sheikh, 2014). The Ladakhis were mostly peons, jamadars, and scale holders. Ladakh was backward in education, and it was the Moravian Mission which established the first English school at Leh in 1875 (Hassnain et al., 1977). In Kargil, the first government school was opened in 1901 and was upgraded into a middle school in 1930 (Sheikh, 2014).

Moreover, the settlement situated on the right bank of the Suru was joined to the unpretentious bazaar on the opposite side by an old sagging bridge (Filippi, 1931), and Bellew mentioned it as the most rickety bridge (Bellew, 1875). As every bridge (zamba) on the Suru River, it had been destroyed by floods in 1894 (Adair & Godfrey, 1899). All forms of communication were entirely disrupted, as postal services and other links between Leh and Central Asia were completely cut off due to the destruction of bridges (Adair & Duke, 1903; Godfrey, 1899). In response, a new road and a suspension bridge were constructed approximately 3.2 km upstream from the old bridge (zamba), with the work completed in January 1902 (Duke, 1903). This newly built Kargil bridge was not only more robust but also significantly larger than the earlier structure crossing the Suru River (Adair & Godfrey, 1899). Meanwhile, the primary bridge over the Indus River, designed, aligned, and initiated in 1893, was completed by early summer 1894 (Adair & Godfrey, 1899). In Ladakh, two main types of bridges were in use: the wooden bridge (Shingzam), typically constructed with large poplar beams, and the suspension bridge (chugzam), made from two thick ropes of twisted birch twigs, roughly the thickness of a man's arm (Cunningham, 1854; Gazetteer of Kashmir and Ladák, 1890).

5.2 Urban Population Change

In Ladakh, the urban population of urban centres is increasing unprecedentedly due to rapid urbanisation. The urban population in Kargil town increased nearly tenfold, rising from 1,681 in 1961 to 16,338 in 2011 (Fig. 5.2). The urbanisation processes, particularly natural population growth, have been the dominant factors, combined with rural-to-urban migration. The town's population experienced a sharp rise in decadal growth rate, increasing from 42.2% in 1971 to 75.4% in 2001, followed by a decrease to 53.3% in 2011. Kargil town's urban population is projected to grow to approximately 29,767 by 2027 and further increase to around 34,232 by 2032 (Town Planning Organisation, 2018). Between 1961 and 2011, Kargil's urbanisation level increased from 3.7 to 11.6%, with intermediate increases to 5.3% in 1981 and 8.9% in 2001—amounting to a rise of 7.9 percentage points.

While Kargil town experienced rapid population growth, the rural population of Kargil district also grew significantly, rising from 43,383 in 1961 to 124,464 in 2011, as per the most recent Census of India. As a result of this population growth, the demand of proper infrastructure, need for social services, and pressure on local

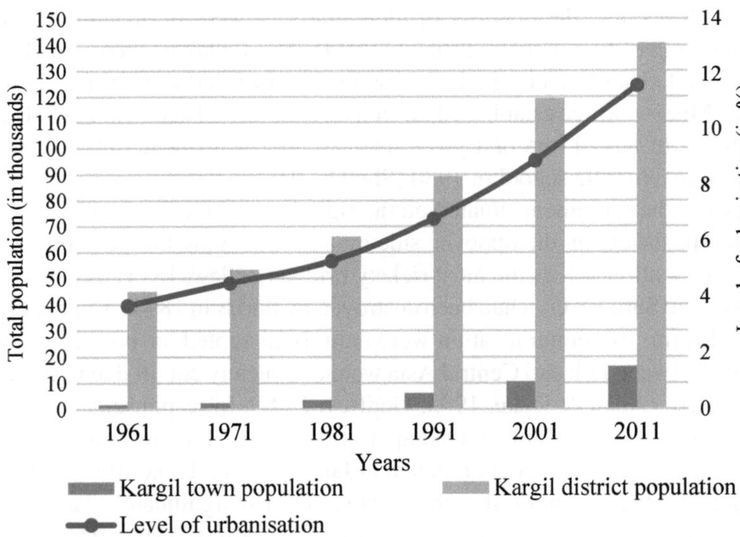

Fig. 5.2 Total town population, district population, and level of urbanisation in Kargil, Ladakh. *Sources* Census of India (1961, 1971, 1981, 2001, 2011a, 2011b, 2011c); the 1991 census was not conducted in Jammu & Kashmir; therefore, the population figures for Kargil town in 1991 were interpolated by the author using the linear projection method

Table 5.1 Population growth trends and decadal growth rates

Years	Kargil town population	Town decadal growth rate (%)	Percentage of urban population (%)	District population	District decadal growth rate (%)	Ladakh population
1961	1,681	–	3.7	45,064	–	88,651
1971	2,390	42.2	4.5	53,400	18.5	105,291
1981	3,527	47.5	5.3	65,992	23.6	134,372
1991	6,074	72.1	6.8	89,334	35.4	179,410
2001	10,657	75.4	8.9	119,307	33.6	236,539
2011	16,338	53.3	11.6	140,802	18	274,289

Sources Census of India (1961, 1971, 1981, 2001, 2011a), Hussain et al. (2023a)

resources have increased in both rural and urban areas. Furthermore, Kargil district's total population tripled from 45,064 in 1961 to 140,802 in 2011, with the decadal growth rate rising from 18.5% in 1971 to 35.4% in 1991 and then falling to 18% in 2011 (Table 5.1).

5.3 Comparison of Blocks Population

From Fig. 5.3, it can be observed that the population of Zanskar block increased significantly from 6,886 in 1971 to 11,653 in 2011. The population of Drass block grew fourfold, from 5,428 in 1971 to 21,988 in 2011, while the population of Sankoo block increased from 11,884 in 1971 to 17,735 in 2011. The population of Kargil block tripled, growing from 13,694 to 41,512 over the same period (Table 5.2). The growth rate of population is significantly higher in Zanskar, Drass, Sankoo, and Kargil blocks compared to the other blocks of Kargil district. The population growth rate in Kargil district was higher compared to Leh district as a result of faster growth in the rural population (Goodall, 2004).

5.4 Population Size of Settlements

The population size of settlements in Kargil district, Ladakh, experienced significant changes between 1981 and 2011, as depicted in Fig. 5.4. Population growth in these settlements can be linked to better road connectivity, the expansion of tourism, increased economic opportunities, and enhanced local services. These factors have made the areas more accessible and attractive for the growth of settlements. Additionally, population growth between 1981 and 2011 was primarily driven by natural

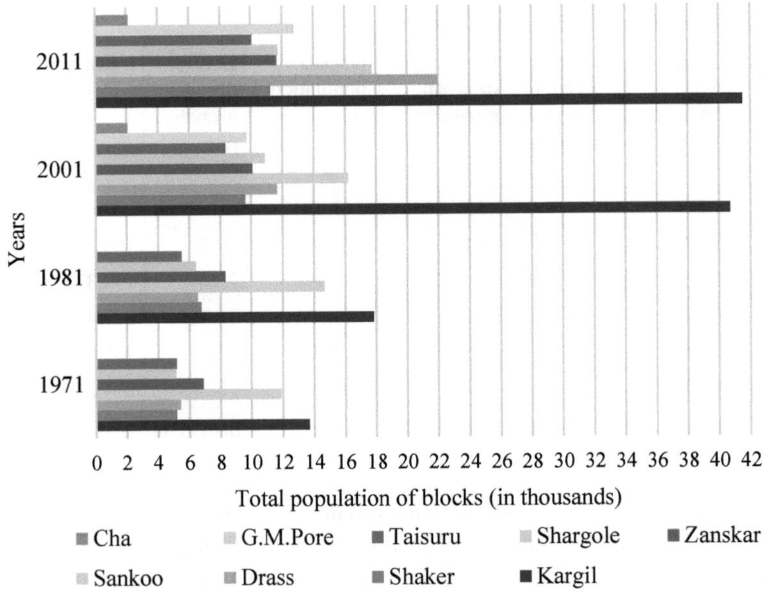

Fig. 5.3 Block-wise total population of Padum, Sankoo, Drass, and Kargil in Kargil district, Ladakh. *Sources* Census of India (1971, 1981, 2001, 2011c)

Table 5.2 Block-wise total population of Padum, Sankoo, Drass, and Kargil in Kargil district

Blocks population	1971	1981	2001	2011
Kargil	13,694	17,813	40,718	41,512
Shaker	5184	6775	9606	11,233
Drass	5428	6544	11,662	21,988
Sankoo	11,884	14,648	16,198	17,735
Zanskar	6886	8307	10,090	11,653
Shargole	5141	6404	10,893	11,728
Taisuru	5183	5501	8345	10,059
G. M. Pore	–	–	9716	12,754
Cha	–	–	2079	2140

Sources Census of India (1971, 1981, 2001, 2011b)

increase, resulting from high birth rates and a decline in mortality rates due to advancements in health care and improved living conditions. The government's initiatives supporting rural livelihoods, improved road connectivity, and economic diversification have contributed to sustaining populations in these settlements. For instance, the development of roads such as NH-1, Kargil-Zanskar highway (NH-301), and the Nimmu-Padum-Darcha (NPD) has significantly transformed the region, enhanced trade, and improved access to healthcare, education, and other services. However, the improved accessibility provided by these roads has also placed considerable pressure on natural resources, leading to environmental degradation in the region.

The rural populations of several key villages experienced notable growth between 1981 and 2011. Pashkyum's population increased from 1,609 to 3,419, while Minjee's grew from 1,023 to 2,331. Akchamal's population increased from 947 to 1,909, and Yourbaltak's from 1,107 to 2,177. Similarly, Apati's population rose from 594 to 1,245, Silmo's from 851 to 1,627, and Hardras's from 653 to 1,480. Additionally,

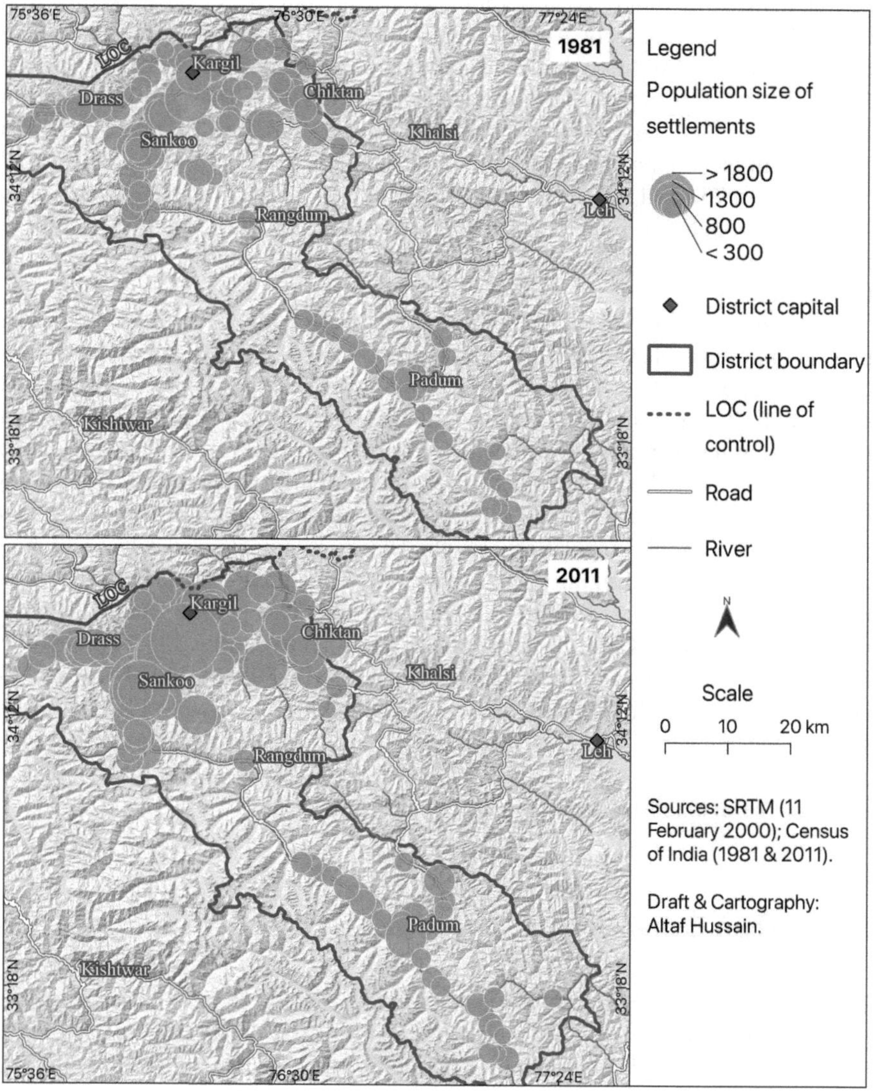

Fig. 5.4 Population size of settlements from 1981 to 2011 in Kargil district, Ladakh

Saleskote saw a jump from 1,051 to 3,605, Trespone's from 1,424 to 3,332, Tambis's from 1,169 to 2,442, and Gund Mangalpur's from 929 to 1,894 during the same period (Fig. 5.4).

In the Zanskar Valley, the population of Karshah grew from 177 to 964, Zangla from 345 to 954, Ating from to 366 to 605, Phey from 332 to 509, and Ofti Pipiting from 286 to 941 between 1981 and 2011. In the Drass Valley, the population of Bhimbat increased from 569 to 1,393, Chokiyal from 398 to 965, Goshan from 622 to 1,485, Kharbu from 454 to 1,074, and Kaksar from 377 to 958 during the same period. In the Suru Valley, Taisuru's population rose from 283 to 589, Parkachik from 642 to 1,138, Youljuk from 701 to 1,196, Namsuru from 496 to 905, and Panikhar from 156 to 325. Additionally, Lankarchey's population increased from 1,236 to 2,425, Faroona from 442 to 893, Sangrah from 657 to 1,183, Thasgam Thevena from 1,318 to 2,567, and Barsoo from 927 to 1,602 between 1981 and 2011. In the Chiktan Valley, Chiktan's population rose from 694 to 1,285, Shakar from 1,256 to 1,922, Samrah from 758 to 1,441, and Hagnis from 886 to 1,624. In the Shargole Valley, Mulbekh's population increased from 915 to 1,393, Wakha from 1,007 to 1,773, Shargole from 258 to 486, and Skambo from 317 to 623 during the same period (Table 5.3).

Leh and Kargil are the two major urban centres in Ladakh. These urban centres play an important role for the rural areas in terms of educational, administrative, and economic opportunities. The hinterland hamlets and villages depend on these two towns for basic needs and other services. Kargil is the only major town in the district, but it now has three urbanised settlements with urban characteristics: Padum, Sankoo, and Drass, all located within Kargil district. The settlements are mostly located along the Indus River and its tributaries, with many developing along the roads, which are characterised by rapid population growth. These settlements also rely on melting glaciers and snow for drinking water and irrigation of agricultural fields. In addition, rural households increased from 10,529 to 16,147 between 1981 and 2011, while urban households in Kargil district grew nearly threefold from 735 to 2,191 over the same period (Census of India, 2011b; Khan, 1981a, 1981b). In Leh district, the number of rural households increased from 12,638 to 14,905 between 1981 and 2011, whereas the number of urban households rose from 2,162 to 7,004 between 1981 and 2011 (Census of India, 2011b; Khan, 1981b). The total population of Leh district increased from 68,380 to 133,487 between the same period.

5.5 Urban Expansion

The process of urbanisation is characterised by a natural increase in urban population, rural-to-urban migration, and the transformation of urban structure. These factors have collectively contributed to the significant growth of buildings and urban infrastructure. The old local bazaar forms the nucleus of the town, around which initial urban development took place, resulting in the highest building density in the

Table 5.3 Total population of the selected villages in Kargil district, Ladakh

Villages	Total population in 1981	Total population in 2011
Abran	228	418
Akchamal	947	1909
Apati	594	1245
Ating	366	605
Barsoo	927	1602
Bartoo	348	679
Bhimbat	569	1393
Bodhkharbu	704	970
Chiktan	694	1285
Chokiyal	398	965
Choskore	1896	3430
Darchiks	305	624
Faroona	442	893
Gindyal	244	676
Goshan	622	1485
Gund Mangalpur	929	1894
Hagnis	886	1624
Hardas	653	1480
Holiyal	308	615
Kaksar	377	958
Kargyak	195	429
Karit	246	526
Karkit	932	2373
Karshah	177	964
Khandi	748	1456
Kharbu	454	1074
Lankarchey	1236	2425
Lotsum	217	405
Matayan	287	500
Minjee	1023	2331
Mulbekh	915	1393
Muradbagh	332	691
Namsuru	496	905
Ofti pipiting	286	941
Pandrass	382	731
Parkachik	642	1138
Pashkyum	1609	3419

(continued)

Table 5.3 (continued)

Villages	Total population in 1981	Total population in 2011
Phey	332	509
Purtikchey	585	963
Salapigaipak	295	701
Saleskote	1051	3605
Samrah	758	1441
Sangrah	657	1183
Sanjak	271	416
Shakar	1256	1922
Shargole	258	486
Silmo	851	1627
Skambo	317	623
Stakpa	463	878
Taisuru	283	589
Tambis	1169	2442
Testa	185	460
Thasgam Thevena	1318	2567
Trespone	1424	3332
Stongde Kumi	724	914
Umba	520	844
Wakha	1007	1773
Yourbaltak	1107	2177
Yuljuk	701	1196
Zangla	345	954

Sources Census of India (1981, 2011a, 2011b, 2011c)

area. Over time, the town has gradually expanded beyond the old bazaar to incorporate surrounding neighbourhoods such as Thaskangrong, Changchik, Balti Bazaar, Changrah, Daythang, and Lankore. Owing to the town's topographical constraints, urban growth and spatial expansion have primarily occurred towards Titichumik in the south, Shilikchey in the north-west, Akchamal in the east, and the Kurbathang Plateau in the south-east. The built-up area has evolved organically, characterised by a high density of buildings, narrow roads, streets, and alleys, with limited open spaces. The spatial distribution of the urban fabric predominantly comprises residential, commercial, administrative, public, and religious structures, reflecting a multifunctional urban landscape. The old structure of the local bazaar, with its old-fashioned shops and narrow streets, can be seen in Fig. 5.5.

Kargil town serves as a nucleus for the surrounding settlements. The town centre is Kargil's commercial hub, comprising the old hospital, religious institutions, taxi stands, the old polo ground (Hussaini Park), shops, offices, restaurants, hotels, and

Fig. 5.5 Old local bazaar of Kargil town, Ladakh. Photograph taken by Altaf Hussain, 25 August 2021

bazaars, which generate significant economic opportunities. The town centre has the potential to become a key attraction for tourists and visitors and could serve as a venue for exhibitions, cultural festivals, and religious events. As an administrative capital, Kargil has experienced rapid residential expansion due to rural-to-urban migration and urban population growth.

The Baroo residential colony was established in the 1970s, followed by the development of Poyen and Tumail colonies during the 1990s. In 1987, the Tumail village (yul), comprising forty households, was damaged due to a land subsidence accident in Sodh (Milton, 2013). To compensate for the accident, land was provided to the villagers of Tumail on the Kurbathang Plateau, and Tumail Colony became the first residential settlement there. Following the 1999 Kargil War, several new colonies were established. Haidery Mohalla, a cluster of dwellings, and Silmo emerged on the left side of the Suru River, while on the right side, within the Kurbathang Plateau, settlements such as Andoo, Barchey, Al-Zehra, and Fatima were developed (Fig. 5.6). These colonies have mostly developed over a period of two decades, particularly after the establishment of LAHDC-Kargil in 2003. Most of these settlements are situated on steep slopes, making them vulnerable to natural hazards and raising concerns about the sustainability of town planning (Hussain et al., 2023b). The demand for developed land has risen significantly to meet the needs of the expanding population in and around Kargil town (Town Planning Organisation, 2018). The pressure on land for infrastructural development has been immense, which has resulted in the construction of buildings on vulnerable slopes. Urban growth and expansion depend on the availability of land, infrastructure development, and judicious land-use planning for the town in the future.

In 2006, the administrative boundaries of Kargil were restructured and extended to incorporate the settlements of Baroo, Poyen, Bagh-e-Khomeini, and Goma Kargil within the jurisdiction of the Kargil MC (Town Planning Organisation, 2018). Until 2010, the urban fabric of the town consisted of both old and newly constructed structures. In the same year, a bypass was completed, which alleviated heavy military and civilian truck traffic from the main bazaar (Rizvi, 2012). The urban landscape of Kargil is rapidly expanding to accommodate the growing population and increasing demand for infrastructure, despite the constraints posed by scarce available space. The central old town, Baroo Colony, and Poyen Colony have already reached saturation due to unplanned construction of buildings and infrastructure (Hussain et al., 2023b). Unplanned development has led to the collapse of houses, particularly those

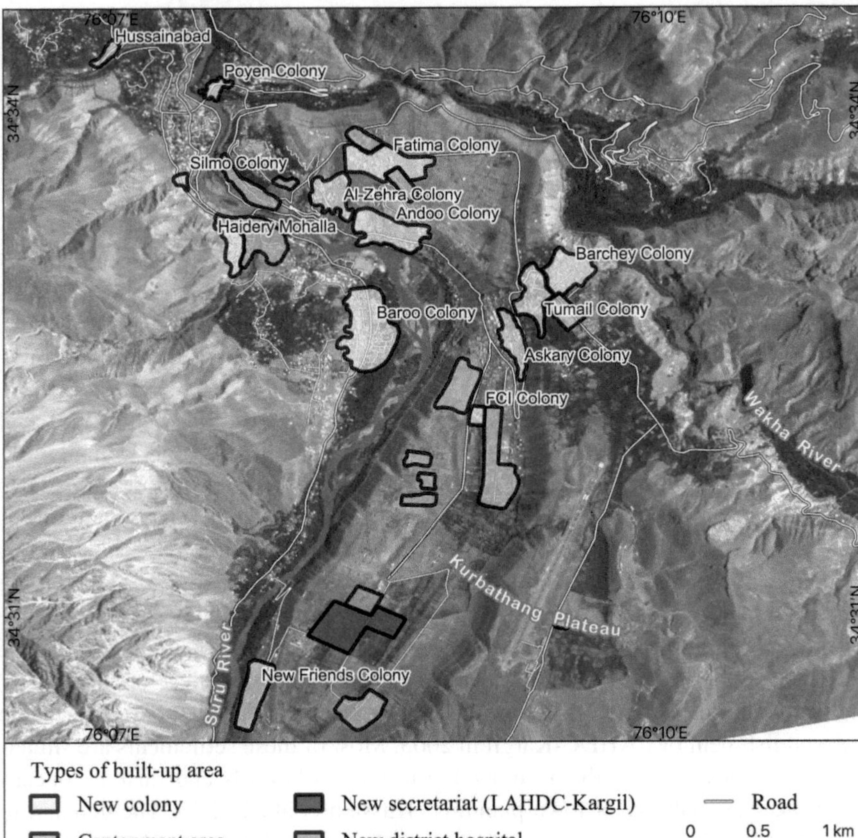

Fig. 5.6 Development of new residential settlements and army areas. *Sources* PlanetScope (8 June 2020), OpenStreetMap (OSM). *Draft and cartography* Altaf Hussain

newly built on steep slopes, resulting in unsafe construction practices (Fig. 5.7). The lack of proper urban planning, combined with the rugged topography, has further exacerbated the vulnerability of these structures to natural hazards like landslides and earthquakes. These factors raise serious concerns about the sustainability of current development patterns in this high-altitude town. Furthermore, the construction of these new buildings is taking place without compliance with building codes and regulations, resulting in unplanned and disorderly urban growth due to the absence of building bylaws and effective town planning (Town Planning Organisation, 2018). The primary factors behind this trend are rapid population increase and rural-to-urban migration. Rural migrants are particularly attracted to the district headquarters of Kargil by the availability of employment opportunities, administrative offices, the district hospital, commercial centres, army installations, banks, and both public and private educational institutions.

Kurbathang Plateau (Fig. 5.8), located near Kargil town, is considered a strategically valuable piece of land, first occupied by the military in 1962 during the Sino-Indian War. The Kargil War of 1999, along with the continuing threat of future incursions from China and Pakistan, further reinforced the military's hold over the Kurbathang Plateau (Bhan, 2013). Over time, the military established small cantonment boards on both sides of the Iqbal Bridge (Zamba) in Kargil town, contributing to its strategic development. A total of 3,959 kanals[2] (200.3 ha) of land were under military occupation in the Kurbathang Plateau until 2009 (Bhan, 2013). Military settlements have played a key role in the urban expansion of this plateau. Since the development of Tumail, the area has seen notable growth in residential colonies, along with the emergence of administrative and commercial structures, such as the newly constructed LAHDC-Kargil Secretariat and the district hospital. LAHDC-Kargil has persistently demanded the reclamation of suitable areas from the military to develop and expand the township on the Kurbathang Plateau. The urban planning challenges are particularly complex in Kargil town, where rapid urbanisation is overtaking the existing capacities of LAHDC-Kargil to manage spatial growth and expansion. As the town continues to expand beyond its municipal boundaries, robust urban planning becomes increasingly necessary for its sustainable development.

Kargil town has witnessed substantial migration from nearby villages, placing considerable strain on its land resources and infrastructure (Town Planning Organisation, 2018). The lack of planned development, along with encroachments on agricultural lands, barren areas, riverbanks, and hillsides, has contributed to serious urban issues, including traffic congestion, poor solid waste management, and inadequate sanitation and drainage systems (Hussain et al., 2023a). The town's core area is heavily congested due to the concentration of most service facilities in one location. This centralisation creates significant problems, particularly for the town's traffic system, as the high density of activities attracts a large number of vehicles and pedestrians. The absence of alternative service hubs exacerbates the situation, as residents, migrants, and visitors heavily depend on the core area for essential services and amenities. As of now, Kargil town is home to over eight private schools alongside

[2] A kanal is a measure or unit of area and is equal to $^1/_8$ acre or 506 m^2.

Fig. 5.7 Newly constructed houses on steep slope areas in Kargil town. Photograph taken by Altaf Hussain (top: 28 August 2021; bottom: 17 August 2024)

Fig. 5.8 Panoramic view of the landscape of the Kurbathang Plateau. The plateau is situated on the right side of the Suru River, near Kargil town. Photograph taken by Altaf Hussain, 8 September 2022

government public schools. There was a significant expansion in banking services, with the number of state and national banks increasing from 13 to 35 between 2009–2010 and 2022–2023 (District Statistical & Evaluation Office Kargil, 2012, 2022). In the Kargil district, the number of shops increased from 1,849 to 4,234 between 2008–09 and 2022–23, while the number of commercial establishments rose considerably from 147 to 839 during the same period (District Statistical & Evaluation Office Kargil, 2012, 2022). Additionally, there was a sharp rise in small-scale industrial units from 14 to 756 between 2014–15 and 2021–22 (District Statistical & Evaluation Office Kargil, 2016, 2021).

5.6 Built-Up Area Change

A comparative analysis of built-up area expansion reveals notable differences between Kargil town and the urbanised settlements of Padum, Sankoo, and Drass (Fig. 5.9). In Padum, the built-up area expanded more than eightfold, from 10 hectares (ha) in 1980 to 82 ha in 2021. Sankoo experienced a more than sixfold increase, with its built-up area growing from 10 ha in 1971 to 65 ha in 2021. Drass also witnessed substantial growth, with the built-up area rising more than eightfold from 12 ha in 1965 to 85 ha in 2021. The most significant change occurred in Kargil town, where the built-up area increased more than ninefold, from 25 ha in 1965 to 230 ha in 2020. This clearly indicates that Kargil town experienced a much larger expansion in built-up area compared to the other urbanised settlements in Kargil district. Leh town, the capital of Ladakh, also witnessed significant urban growth, with its built-up area expanding from 36 ha in 1969 to 196 ha in 2017 (Dame et al., 2019).

Moreover, a comparison of built-up areas in Kargil town and the urbanised settlements per year reveals notable trends in urban expansion. The built-up area increased by 1.76 ha per year in Padum between 1980 and 2021, by 1.10 ha per year in Sankoo between 1971 and 2021, by 1.30 ha per year in Drass between 1965 and 2021, and by 3.73 ha per year in Kargil between 1965 and 2020. The urbanisation processes, combined with the increasing population, have significantly altered the LULC dynamics, particularly in Kargil town and the urbanised settlements of Padum, Drass, and Sankoo.

The rapid urbanisation, combined with demographic growth, has resulted in the construction of buildings and infrastructure development. The town, once dominated by irrigated land and vegetation, has transformed into a landscape of residential, commercial, administrative, and other buildings. Most of the old buildings were made of sun-dried clay bricks (pagbu), but these have been replaced by modern buildings using imported materials. For instance, traditional houses were typically double-storeyed with mud-based construction. The plinths and foundations were built with stone. The walls were made of either sun-dried bricks (pagbu) or stones plastered with clay (kalaq). The roofs were flat and supported by poplar beams. The gaps between the wooden logs of the beams were covered with twigs from willow (malchang) or poplar (zberpa) trees. Finally, the rooftops were plastered with a thick layer of mud

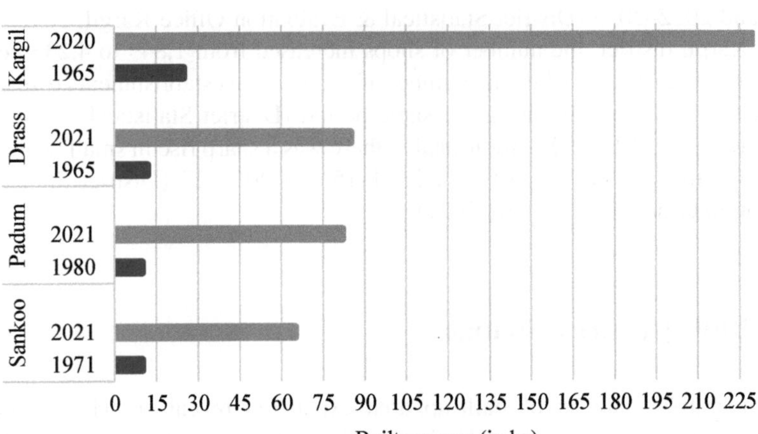

Fig. 5.9 Built-up area changes in Kargil town and the urbanised settlements of Sankoo, Padum, and Drass. *Sources* Satellite images: Kargil—Corona (8 October 1965), PlanetScope (8 June 2020); Drass—Corona (8 October 1965), PlanetScope (18 July 2021); Padum—Corona (11 September 1980), PlanetScope (24 July 2021); Sankoo—Corona (28 September 1971), PlanetScope (12 August 2021)

mixed with dry grass or straw. The modern houses are built with sun-dried bricks, cement bricks, iron bars, stones, and wood as the main construction materials. Poplar (zberpa) beams and willow (malchang) twigs are still used in constructing the roofs, which are layered with clay (kalaq). These construction materials are predominantly used in residential, administrative, and commercial buildings. Moreover, unplanned development not only causes problems with land use but also disrupts the sustainable ecosystem of the town.

The old town, including the bazaar, is characterised by a high built-up density, with narrow streets and compact buildings. Rapid urban growth has led to a significant decrease in agricultural land, as vast areas surrounding the old town have been converted into built-up spaces. This growth has also extended to the Kurbathang Plateau, where the intensification of building and construction activities is evident. Together, these changes highlight the ongoing urban expansion and its significant impact on the surrounding landscape. Rural-to-urban migration is increasingly contributing to urban growth, as more migrants move from rural areas in search of employment, healthcare, and educational opportunities. Many of these migrants tend to settle in Kargil, the administrative capital. As a result, new houses are being built in the available space, particularly in Kurbathang, leading to a significant reduction in open space. The absence of effective land-use planning and the challenge of managing the rapid change in the urban landscape are major issues for LAHDC-Kargil and the UT of Ladakh. Rapid urbanisation has significantly increased the demand for municipal infrastructure; however, the current quality of municipal infrastructure is insufficient to meet the needs of both the existing urban population and incoming migrants, posing serious concerns for the long-term sustainability of Kargil town. The spatial transformation is associated with the rapid urbanisation, which is occurring haphazardly. The repeat photography of Baroo and the old town of Kargil, shown in (Figs. 5.10 and 5.11), provides a clear visual representation of the urban landscape's transformation. These images effectively highlight the significant changes over time, from 1927–1928 to 2024, illustrating the impact of urbanisation on both the natural surroundings and the built environment. Similarly, the changes on the Kurbathang Plateau are evident from the repeat photography conducted between 1977 and 2025 (Fig. 5.12).

In Ladakh, rapid urbanisation has emerged as a major driving force behind the transformation of settlements. A key factor contributing to this shift is the development of road infrastructure, along with the expansion of the tourism sector. These changes have accelerated modifications in LULC, fundamentally reshaping the physical landscape. As a result, built-up areas have increased significantly, often expanding in an unplanned and poorly regulated manner, posing challenges for sustainable urban development and environmental management. Additionally, in the absence of effective mechanisms among Development Authorities, Urban Local Bodies (ULBs), and the Housing and Urban Development Department (H&UDD), the Ladakh region has suffered due to a lack of coordination and cooperation between local and central authorities in the urban planning and provision of infrastructure services. There are only two municipal committees, and more municipalities need to be established constitutionally, without considering the population criteria, to

Fig. 5.10 Repeat photography of Baroo in Kargil town. The photographs illustrate the changes in the urban structure between 1927–1928 and 2024. The intensive growth of buildings over time is evident due to urbanisation processes. *Sources* **a** Helmut De Terra, photograph taken between 1927 and 1928, American Geographical Society Library, University of Wisconsin-Milwaukee Libraries; **b** photograph taken by Altaf Hussain, 17 August 2024

Fig. 5.11 Repeat photography of Kargil's old town. The photographs depict the urban structure of the old town of Kargil on the left side of the Suru River, where the conversion of irrigated land and vegetation into built-up areas can be observed. *Sources* **c** Helmut De Terra, photograph taken between 1927 and 1928, American Geographical Society Library, University of Wisconsin-Milwaukee Libraries; **d** photograph taken by Altaf Hussain, 17 August 2024

Fig. 5.12 Repeat photography of the Kurbathang Plateau. The images show the conversion of barren land into vegetation and built-up areas, highlighting the substantial changes in the landscape over the years. Photograph taken by Monique Fort, 1977 (top); photograph taken by Altaf Hussain, 12 June 2025 (bottom)

address the challenges and issues. The absence of planned urban development and insufficient enforcement of regulations have contributed to rapid and unregulated urban sprawl, leading to major transformations in the region's urban footprint. If urban development continues without proper urban planning in existing urban areas, there will be unprecedented stress on environmental and socio-economic resources. A long-term blueprint or vision is urgently needed to foster growth, inclusivity, and sustainability in urban centres, particularly in the district capitals of Leh and Kargil, to ensure the sustainable development in Ladakh.

5.7 Tourism Sector

The tourism sector has been one of the major drivers of urbanisation in Ladakh. The development of Kargil's tourism sector commenced gradually following the opening of Ladakh to tourism in 1974. Tourist arrivals rose from 4,325 in 1976 to 17,501 in 1989. However, a sharp decline followed between 1990 (1,134) and 2003 (4,244), as Kargil was severely impacted by the Kashmir conflict (Grist, 2008), which eventually led to the Kargil War in 1999. After 2003, tourism witnessed exponential growth, particularly among domestic visitors, although a slight decline was observed in 2013 and 2014 (Fig. 5.13). Tourist numbers rose significantly, reaching 28,756 in 2010 and 109,558 by 2018. However, between 2019 and 2021, the number of visitors declined due to the bifurcation of Jammu and Kashmir in 2019, followed by international travel restrictions and nationwide COVID-19 lockdowns in India during 2020 and 2021. After the lockdown, tourist arrivals in Kargil saw a significant increase, more than tripling between 2018 and 2024, with a total of 324,647 visitors recorded in 2024. This increase was driven by the easing of travel restrictions, improved infrastructure, and renewed interest in the region as a tourist destination. Despite this growth, the number of international tourists remained considerably lower than domestic visitors.

The tourism sector is highly unpredictable, influenced by seasonal factors, accessibility challenges, and its reliance on geopolitical stability. Geopolitical conflicts, particularly tensions along the LOC and border disputes, have a substantial impact on tourism, often discouraging international travellers. Furthermore, inadequate infrastructure, limited promotion of Kargil as a global tourist destination, and the perception of the region as a conflict zone further contribute to the disparity in international arrivals. Kargil is often viewed as a transit point, with tourists typically staying for a day or more, depending on their plans, before continuing their journey to Leh, which offers more attractions and is considered the primary destination in the region.

The rise in tourist arrivals led to a significant increase in hospitality infrastructure. The number of hotels increased more than twofold, rising from 17 in 2015–2016 to 49 in 2023–2024, while guesthouses saw a modest rise from 70 to 115 during the same period (Fig. 5.14). The number of travel agencies also grew, increasing from 40 in 2015–2016 to 87 in 2023–2024. The tourist multiplied from 4,325 in 1976 to 225,543 in 2022, which led to massive increase of hotels, guesthouses, and travel agencies (Hussain et al., 2023a). The number of hotels and restaurants has

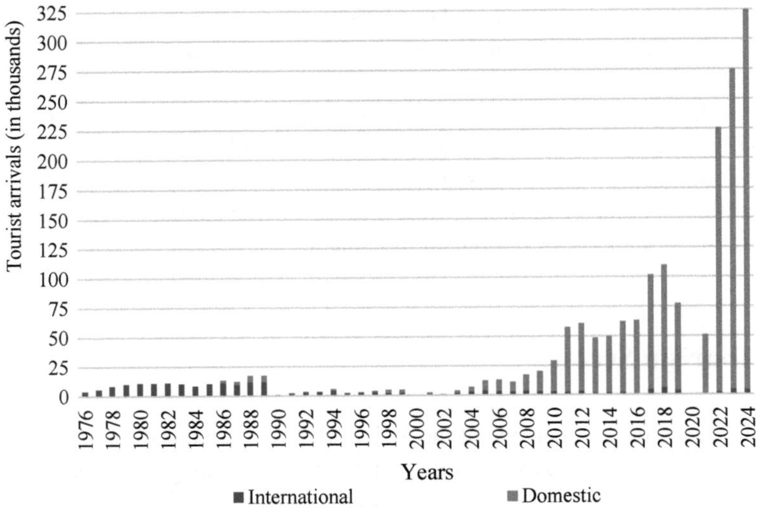

Fig. 5.13 Domestic and international tourist arrivals in Kargil, Ladakh, from 1976 to 2024. *Sources* District Tourism Department of Kargil, UT of Ladakh

risen from 344 to 770 between 2008–09 and 2022–23 in Kargil district (District Statistical & Evaluation Office Kargil, 2012, 2023). The tourism sector in Kargil is rapidly expanding due to the rise in tourism activities, commercial interactions, and improved road connectivity. This has become a major force in reshaping and transforming the urban landscape of Kargil town.

The number of hotels and guesthouses has grown significantly over the past few decades. In 1974, there were 20 hotels and guesthouses in Kargil town (Schettler & Schettler, 1989). By 2024, the number of hotels and guesthouses had increased to 164 representing a monumental expansion of lodging facilities in Kargil town (District Statistical & Evaluation Office Kargil, 2024). The spatial distribution of hotels and guesthouses in Kargil town can be observed from 1974 to 2024 (Fig. 5.15). Urbanisation is driven by factors such as growing urban population, rural-to-urban migration, and the rise of mountain tourism, resulting in new hotel constructions and increased tourist arrivals (Hussain et al., 2023b). The burgeoning presence of hotels and guesthouses not only caters to the diverse needs of tourists but also contributes significantly to the local economy through employment generation and infrastructural development. The trajectory of the tourism sector in the coming years will be influenced by various factors including tourism trends, advancements in infrastructure, and external variables such as economic stability and geopolitical conditions.

The growth of tourism sector has increased unprecedentedly in the neighbouring capital of Leh. The number of hotels and guesthouses increased from 24 in 1980s to more than 360 by 2012 (Gondhalekar et al., 2015). In addition, tourist arrivals in Leh increased from 28,400 in 2003 to 79,100 in 2010, and further to 323,590 in 2018, primarily due to a sharp rise in domestic visitors (Dame et al., 2019). The growth of

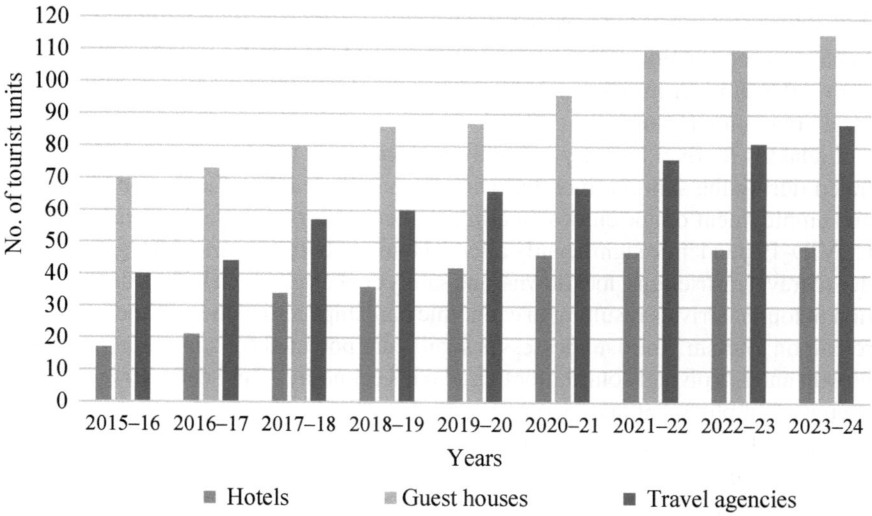

Fig. 5.14 Hotels, guest houses, and travel agencies in Kargil, Ladakh. *Sources* District Tourism Department of Kargil, UT of Ladakh

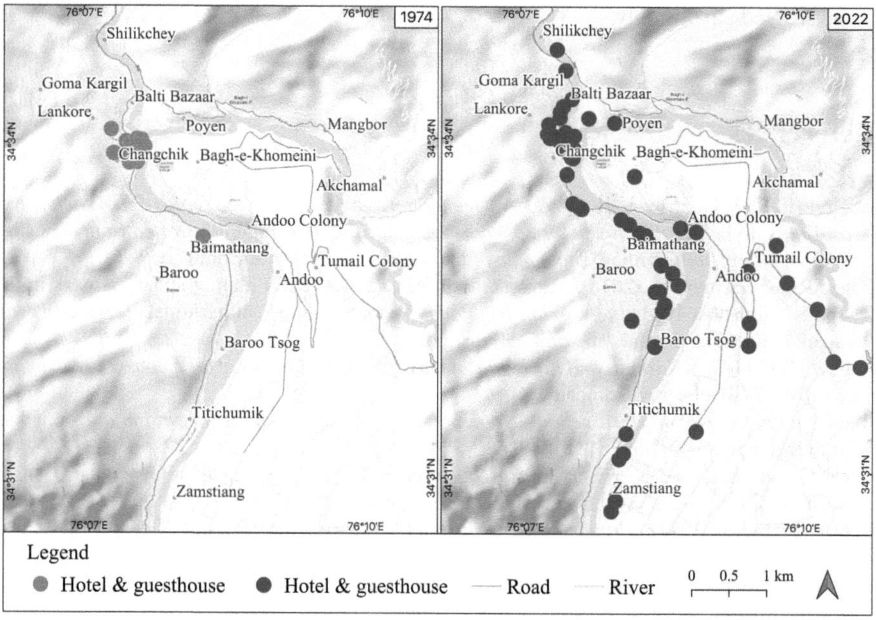

Fig. 5.15 Spatial distribution of hotels and guesthouses in Kargil town from 1974 to 2024 in Ladakh. *Sources* the data of 1974 was taken from Rolf and Margret Schettler book (Schettler & Schettler, 1989) during their visit in Ladakh in 1974; the data of 2024 was compiled from District Tourism Department of Kargil, Ladakh, Google Earth Information, and OSM

the tourism sector has led to increased labour migration, especially from Nepal and India's lowland regions. Tourist arrivals in Ladakh are predominantly concentrated during the summer months and offering seasonal employment opportunities to the local population (Dame et al., 2019).

Ladakh's tourism industry is highly seasonal with most tourist arrivals concentrated during the summer months. This seasonality leads to fluctuations in income and employment opportunities creating challenges for sustainable livelihoods. The COVID-19 pandemic significantly affected tourism both in Leh and Kargil in Ladakh due to travel restrictions, lockdowns, and safety concerns, which led to a drastic reduction in tourist arrivals resulting in economic hardships for businesses and individuals reliant on tourism. Tourism possesses significant potential for stimulating economic growth and is a pivotal contributor to Ladakh's economy. This sector serves as a vital source of employment and income for a substantial portion of the local population. The natural landscape, mountain peaks, glaciers, cultural heritage, and history are the main tourism attracting assets of the region, which is attracting domestic and international tourists from Indian states and foreign countries. The rapid increase in tourism has raised concerns regarding its ecological impact on Ladakh's fragile environment recently. Urgent issues such as waste management, water scarcity, pollution, and damage to natural habitats have become pressing challenges that need to be addressed sustainably.

References

Adair, F. E. S., & Godfrey, S. H. (1899). *A Summer in High Asia: Being a record of sport and travel in Baltistan and Ladakh, with an appendix on Central Asian trade, illustrated from drawings by the author, photographs, and a map of the route.* W. Thacker & Company.

Bates, R., & Harman, N. (2014). The lost world of Ladakh—Early photographic journeys in the Indian Himalaya 1931–1934. *Asian Highlands Perspectives*, 31. http://himalaya.socanth.cam.ac.uk/collections/journals/ahp/pdf/AHP_31.pdf

Bellew, H. W. (1875). *Kashmir and Kashghar: A narrative of the journey of the embassy to Kashghar in 1873–74.* Trübner & Company.

Bhan, M. (2013). *Counterinsurgency, democracy, and the politics of identity in India.* Routledge. https://doi.org/10.4324/9781315889863

Census of India. (1961). *Primary census abstract.* Office of the Registrar General & Census Commissioner, India. https://censusindia.gov.in

Census of India. (1971). *Primary census abstract.* Office of the Registrar General & Census Commissioner, India. https://censusindia.gov.in

Census of India. (1981). *Primary census abstract.* Office of the Registrar General & Census Commissioner, India. https://censusindia.gov.in/

Census of India. (2001). *Primary census abstract.* Office of the Registrar General & Census Commissioner, India. https://censusindia.gov.in/

Census of India. (2011a). *District census handbook Leh (Ladakh): Village and town wise, series-02, part XII-B.* Directorate of Census Operations, Jammu & Kashmir.

Census of India. (2011b). *District census handbook Kargil: Village and town wise, series-02, part XXI-B.* Directorate of Census Operations, Jammu & Kashmir.

Census of India. (2011c). *Primary census abstract.* Office of the Registrar General & Census Commissioner, India. https://censusindia.gov.in/

Cunningham, A. (1854). *Ladák, physical, statistical, and historical: with notes of the surrounding countries*. Wm. H. Allen and Company.

Dame, J., Schmidt, S., Müller, J., & Nüsser, M. (2019). Urbanisation and socio-ecological challenges in high mountain towns: Insights from Leh (Ladakh), India. *Landscape and Urban Planning, 189*, 189–199. https://doi.org/10.1016/j.landurbplan.2019.04.017

District Statistical & Evaluation Office Kargil. (2012). *District Kargil at a Glance*. Directorate of Economics & Statistics, Planning Development & Monitoring Department.

District Statistical & Evaluation Office Kargil. (2016). *Statistical handbook*. Directorate of Economics & Statistics, Planning Development & Monitoring Department.

District Statistical & Evaluation Office Kargil. (2021). *Statistical handbook*. Directorate of Economics & Statistics, Planning Development & Monitoring Department.

District Statistical & Evaluation Office Kargil. (2022). *Statistical handbook*. Directorate of Economics & Statistics, Planning Development & Monitoring Department.

District Statistical & Evaluation Office Kargil. (2023). *Statistical handbook*. Directorate of Economics & Statistics, Planning Development & Monitoring Department.

District Statistical & Evaluation Office Kargil. (2024). *Statistical handbook*. Directorate of Economics & Statistics, Planning Development & Monitoring Department.

Drew, F. (1875). *The Jummoo and Kashmir territories: A geographical account*. Edward Stanford.

Duke, J. (1903). *Kashmir and Jammu: A guide for visitors*. Thacker, Spink & Co.

Duncan, J. E. (1906). *A summer ride through western Tibet*. Smith, Elder & Company.

Filippi, F. D. (1931). *The Italian expedition to the Himalaya, Karakoram and Eastern Turkestan (1913–1914)*. Edward Arnold and Co.

Francke, A. H. (1926). F. W. Thomas (Ed.), *Antiquities of Indian Tibet: Part II-the chronicles of Ladakh and minor chronicles, texts and translations, with notes and maps* (Vol. L). Superintendent Government Printing, India.

Gazetteer of Kashmir and Ladák. (1890). *Together with routes in the territories of the Maharája of Jamú and Kashmír; compiled under the direction of the quarter master general in India in the intelligence branch*. Superintendent of Government Printing, India.

Gompertz, M. L. A. (1926). *The road to Lamaland: Impressions of a journey to Western Tibet*. Hodder & Stoughton Limited.

Gondhalekar, D., Nussbaum, S., Akhtar, A., & Kebschull, J. (2015). Planning under uncertainty: Climate change, water scarcity and health issues in Leh town, Ladakh, India. In W. Leal Filho, & V. Sümer (Eds.), *Sustainable water use and management* (pp. 293–312). Springer International Publishing. https://doi.org/10.1007/978-3-319-12394-3_16

Goodall, S. K. (2004). Rural-to-urban migration and urbanization in Leh, Ladakh: A case study of three nomadic pastoral communities. *Mountain Research and Development, 24*(3), 220–227. https://doi.org/10.1659/0276-4741(2004)024[0220:RMAUIL]2.0.CO;2

Grist, N. (2008). Urbanisation in Kargil and its effects in the Suru Valley. In M. van Beek, & F. Pirie (Eds.), *Modern Ladakh: Anthropological perspectives on continuity and change*. Brill. https://doi.org/10.1163/ej.9789004167131.i-313

Hassnain, F. M., Oki, M., & Sumi, T. D. (1977). *Ladakh: The Moonland* (2nd ed.). Light & Life Publishers.

Hussain, A., Schmidt, S., & Nüsser, M. (2023a). Dynamics of mountain urbanisation: Evidence from the Trans-Himalayan town of Kargil, Ladakh, India. *Land, 12*(4), 920. https://doi.org/10.3390/land12040920

Hussain, A., Schmidt, S., & Nüsser, M. (2023b). *Urban landscape change in the Trans-Himalayan town of Kargil, Ladakh, India*. https://doi.org/10.5194/egusphere-egu23-2663

Khan, K. M. A., Bray, J., Devers, Q., & Vernier, M. (2014). Chigtan castle and mosque: A preliminary historical and architectural analysis. In E. F. L. Bue, & J. Bray (Eds.), *Art and architecture in Ladakh: Cross-cultural transmissions in the Himalayas and Karakoram* (Vol. 35). Brill.

Khan, A. H. (1981a). *Census of India: Town directory, series-8, part XA*. Director of Census Operations, Jammu & Kashmir.

Khan, A. H. (1981b). *District census handbook, Leh (Ladakh) district: Village and town directory, series-8, part XIII-A & B*. Director of Census Operations, Jammu & Kashmir. https://censusindia.gov.in/nada/index.php/catalog/29355

Khan, H. (1992). *Mukhtasar Tareekh Jammu Wa Kashmir*. Jay Kay Book House.

Milton, N. (2013). *Kargil: An abode of Mountain Tamers*. Har-Anand Publications Pvt Ltd.

Moorcroft, W., & Trebeck, G. (1841). *Travels in the Himalayan provinces of Hindustan and the Panjab; in Ladakh and Kashmir; in Peshawar, Kabul, Kunduz and Bokhara; from 1819 to 1825* (Vol. II). John Murray.

Neve, A. (1918). *The tourist's guide to Kashmir, Ladakh, Skardo, & C.* (11th ed.). The Civil & Military Gazette Press.

Petech, L. (1977). *The Kingdom of Ladakh C. 950–1842 A.D.* Istituto Italiano per il Medio ed Estremo Oriente.

Rizvi, J. (2012). *Ladakh: Crossroads of High Asia* (3rd ed). Oxford University Press.

Schettler, M., & Schettler, R. (1989). *Kashmir, Ladakh & Zanskar: A travel survival kit* (3rd ed). Lonely Planet.

Sheikh, A. G. (2014). *Reflections on Ladakh, Tibet and Central Asia*. Inamullah Abdulmatin.

Town Planning Organisation. (2018). *Master Plan of Kargil* (pp. 1–135). https://kargil.nic.in/document/draft-master-plan-kargil/

Vohra, R. (1989). Early history of Ladakh: Mythic Lore and Fabulation. In H. Osmaston, & P. Denwood (Eds.), *Proceedings of the fourth and fifth International Colloquia on Ladakh*. Motilal Banarsidass.

von Schlagintweit, H., Schlagintweit, A., & Schlagintweit, R. (1861). *Results of a scientific mission to India and High Asia* (Vol. 1). Trubner and Company.

Chapter 6
Urban Infrastructure Challenges

Abstract In recent years, the Indian Trans-Himalayan region has experienced rapid transformation due to urban development. This chapter specifically focuses on urban development and its implications in Kargil, the second largest town in Ladakh. The urban development schemes implemented by both state and central governments have resulted in significant changes to the existing urban infrastructure. These initiatives have been instrumental in reshaping the urban landscape of Kargil town. As an administrative centre, Kargil faces several critical urban challenges such as municipal solid waste management, water supply, public toilets, deep drains, and urban mobility. These urban challenges have primarily arisen due to inadequate governance structures, unplanned development, and a lack of comprehensive urban planning. The chapter aims to analyse the ongoing efforts by local (LAHDC-Kargil) and central government urban programmes that aim to address these urban challenges and issues within the UT of Ladakh. This comprehensive analysis seeks to highlight the complexities of urban development initiatives in high-altitude regions, emphasising the importance of urban planning and governance frameworks for sustainable urban growth and development.

Keywords Urban development · Infrastructure · Water supply · Municipal solid waste management · Public toilets · Deep drains · Urban mobility · Challenges · Kargil town · Ladakh

6.1 Town Planning

In India, states began to enact town planning legislation, with the Bombay Town Planning Act of 1915 being the first to be introduced (Ballaney, 2008; National Commission on Urbanisation, 1988). This act was a significant milestone in addressing the challenges posed by rapid urban growth and establishing organised urban development. It empowered local authorities (municipalities) to formulate and execute town planning schemes for both redevelopment and urban development (Ballaney, 2008; Spodek, 2013). In 1915, Patrick Geddes prepared improvement plans for towns like

A. Hussain, *Urbanisation and Development Challenges in Kargil, Ladakh*, Advances in Asian Human-Environmental Research, https://doi.org/10.1007/978-3-032-05667-2_6

Tanjore, Madurai, Lucknow, and Indore. These plans were based on the idea of 'place, work, and folk' and challenged the prevailing official planning approaches rooted in engineering and business conventions (National Commission on Urbanisation, 1988). In 1919, New Delhi was planned and developed by Sir Edwin Lutyens, based on the principles of garden cities (National Commission on Urbanisation, 1988). In 1951, Chandigarh was planned by Le Corbusier and had an important landmark in the evolution of urban planning during post-independence era (National Commission on Urbanisation, 1988). In 1996, the first Urban Development Plan Formulation and Implementation (UDPFI) guidelines were prepared and circulated to all States and UTs of India (NITI Aayog, 2021). Most Master Plans were prepared along the lines of the first Master Plan of Delhi, 1962, prior to the UDPFI guidelines (NITI Aayog, 2021).

The Integrated Development of Small and Medium Towns (IDSMT) was initiated in 1979–80 during the Sixth Plan (IDSMT, 1995). The programme emphasised strengthening small and medium towns as regional growth centres, with the aim of reducing migration from rural areas and smaller towns to large and metropolitan cities (IDSMT, 1995). The Integrated Development of Medium Towns (IDMT) scheme was implemented by ULBs across all towns, including Leh and Kargil, to provide basic civic amenities (Housing & Urban Development Department, 2012). This scheme was executed through Municipal Councils and Committees, with funds channelled via the Directors of ULBs. The Jawaharlal Nehru National Urban Renewal Mission (JNNURM) was one of the major urban schemes launched by Government of India on 3 December 2005. To ensure broader urban development, two additional schemes were introduced alongside JNNURM to target cities and towns that did not qualify as mission cities under the programme. These were Urban Infrastructure Development Scheme for Small and Medium Towns (UIDSSMT) and Integrated Housing and Slum Development Programme (IHSDP) (Housing & Urban Development Department, 2012). JNNURM catalysed investments of approximately $24 billion in urban infrastructure across Indian cities, with the mission approving projects worth $11.2 billion from government-allocated resources (World Bank, 2013). Since 2014, the Government of India has undertaken significant measures to address urban challenges by launching a series of urban development programmes, including the Swachh Bharat Mission (SBM), Atal Mission for Rejuvenation and Urban Transformation (AMRUT), Housing for All or Pradhan Mantri Awas Yojana (PMAY), and the Smart Cities Mission (SCM) (Fig. 6.1).

In Ladakh, Kargil town has experienced unplanned urban development in recent years. The town has faced significant pressures and challenges due to its rapidly growing population, rural-to-urban migration, and the expansion of modern infrastructure. Urban development schemes introduced by the state and central governments have significantly transformed the town's infrastructure and have played a vital role in shaping the urban landscape of Kargil's municipal area. As an administrative centre, it has faced critical urban challenges, including municipal solid waste management, water supply, public toilets, deep drains, and urban mobility. The leading causes of these critical urban challenges have been a lack of governance

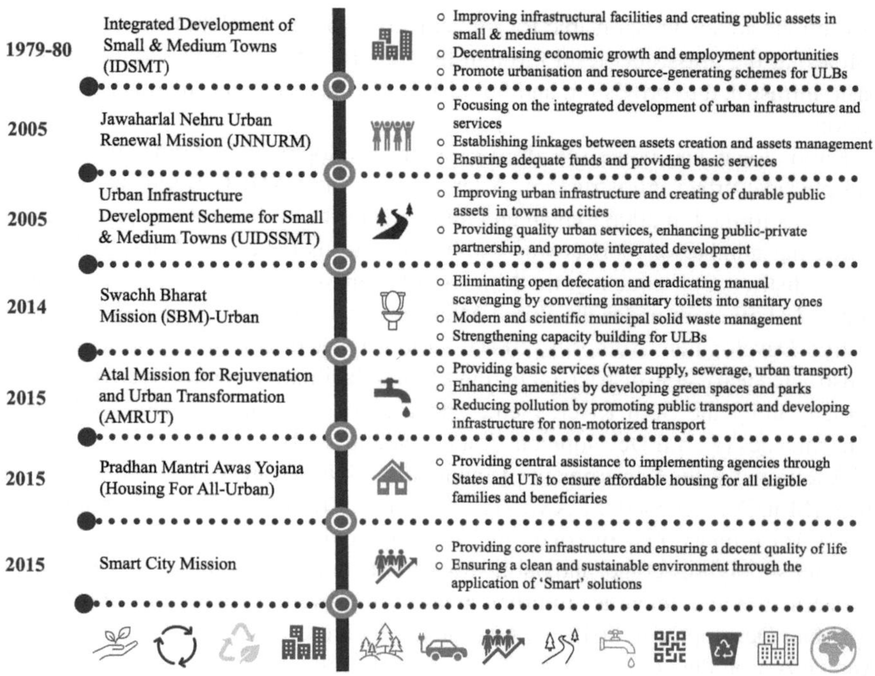

Fig. 6.1 Urban development schemes of Government of India from 1979/80 to 2015. *Draft* Altaf Hussain

structure, unplanned development, and inadequate urban planning. The 74th Constitutional Amendment Act, 1992, empowers municipalities by assigning them responsibilities to function as institutions of self-government, including the preparation of plans for economic development, social justice, and the implementation of development schemes (The Constitution 74th Amendment Act, 1992). In Ladakh, however, municipal governance operates differently due to the establishment of LAHDCs. The LAHDCs Act, first enacted in 1995 and re-enacted in 1997, primarily defines the administrative structure and responsibilities of the LAHDC-Kargil and LAHDC-Leh (The Ladakh Autonomous Hill Development Councils Act, 1997). The Act mentions development programmes and annual or periodic plans, these provisions are broad and administrative rather than spatial, offering little direction for managing urban growth and development. Even ULBs, Development Authorities, and H&UDD lack key planning powers such as zoning, building regulation, and land use control, significantly weakening urban management in the region. The H&UDD has served as the nodal agency for ensuring inclusive and planned growth since Ladakh became a UT on 31 October 2019. Prior to this change, urban development responsibilities were decentralised and managed by the erstwhile Jammu and Kashmir government. The ULBs, Development Authorities, H&UDD, and other related bodies operate under the overall supervision of the LAHDCs and the UT administration of Ladakh, with the LAHDCs focusing on the district-level planning and local governance. These institutions are also responsible for implementing urban development schemes and policies at the grassroots level through their coordinated roles in service delivery, basic infrastructure provision, and broader urban management functions across the region.

Kargil town has witnessed drastic urban development in terms of its social and physical urban infrastructure. After granting the LAHDC-Kargil in 2003, the KDA was formed in 2006 to plan and implement urban infrastructure projects. The increasing number of tourists was one of the reasons for establishing the KDA in Kargil district. Consequently, the Ministry of Tourism, Government of India, sanctioned ₹24.59 crore under the Prime Minister Reconstruction Programme (PMRP) (Kargil Development Authority, 2007). Various types of tourist infrastructure were constructed across the district, including a tourist facilitation centre, riverfront development along the Suru River, picnic spots, tourist bungalows, cafés, and public conveniences (Kargil Development Authority, 2007, 2011).

Additionally, the Government of India, through the Ministry of Urban Development (MoUD), launched the AMRUT on 25 June 2015 (Jammu and Kashmir State, 2017). The primary objective of the mission is to provide essential urban services such as water supply, sewerage, and public transport to households, while also enhancing infrastructure and amenities in urban areas to improve the quality of life (Ministry of Urban Development, 2015). A total of ₹39.92 crore was spent through the State Annual Action Plan of the AMRUT scheme in Kargil town between 2017 and 2020 (Jammu and Kashmir State, 2017). This amount was utilised for the development of green spaces, water supply, sewerage, and septage management in Kargil town. The beautification of urban spaces, such as Bulbul Bagh, Sher Bagh, Indira Park, and Hussaini Park, along with the development of a lift water supply system, was executed by the implementing agencies (PHE and R&B) of LAHDC-Kargil[1] at the grassroots level. The laying of pavement tiles on footpaths/sidewalks for urban beautification was carried out in all wards by the MC under the District Plan of LAHDC-Kargil.

Under AMRUT 2.0 (October 2021), the UT administration of Ladakh has envisaged and allocated an investment of ₹120 crore for the districts of the Ladakh region (All India Radio, 2022). Additionally, the PMAY scheme has been introduced to support and provide affordable housing, addressing housing shortages and improving living conditions of low-income households. This resulted in the construction of 26 houses in 2016–17, 305 houses in 2017–18, 112 houses in 2018–19, 350 houses in 2019–20, and 678 houses in 2020–21 under the PMAY scheme. Each beneficiary received total financial assistance of ₹166,666, while the remaining cost of the house was borne by the beneficiary. The PMAY seeks to alleviate urban housing deficits by providing permanent housing to all eligible urban households, with a focus on economically weaker sections and low- and middle-income groups, as well as slum

[1] The first Master Plan of Kargil was made in 2018 by Feedback Infra Private Ltd. in Ladakh. There are so many flaws in this plan with respect to zoning and land-use planning. A detail master plan is needed to be prepared to address and regulate the urban needs and uncontrolled development of the Kargil town. Land use, building codes, solid waste management, traffic problems, and infrastructure improvement are major issues that need to be dealt with a robust planning framework.

dwellers (Ministry of Housing & Urban Poverty Alleviation, 2015). Urban development efforts in the Leh and Kargil districts primarily focus on ensuring a round-the-clock water supply, establishing sewerage systems, developing public parks, and rejuvenating local water bodies (All India Radio, 2022). In Leh town, urban households, public institutions, and tourist facilities primarily depend on soak pits and septic tanks, which pose significant risks to water quality, especially through the contamination of drinking water sources and groundwater (Gondhalekar & Akhtar, 2015; Müller et al., 2020). To ensure water supply, the urban population has been extracting significant amounts of groundwater from the aquifers of the Indus River using tube wells and borewells (Gondhalekar et al., 2015; Lahmouri et al., 2019). The Jal Jeevan Mission (JJM) is being implemented in Ladakh to ensure the provision of safe and sufficient drinking water to households in both rural and urban areas.

6.2 Urban Water Supply

Kargil town has undergone water supply shortages in residential areas, commercial areas, and irrigated fields. The demand for water supply has increased due to population growth, rural-to-urban migration, expansion of the tourism sector, and rapid urbanisation. Additionally, the melting of snow and glaciers has contributed to the water shortage. The PHE department of LAHDC-Kargil is the sole government agency providing water to the urban population through piped water, pumps/borewells, and water tankers. In Kargil town, providing access to clean water supply has been a significant challenge, particularly for urban households. These households rely on piped water systems, tankers, and borewell drilling for their drinking water needs. According to Mirza Mehdi, Assistant Executive Engineer of the PHE department, LAHDC-Kargil (interview conducted on 17 July 2023), there has been an insufficient water supply in Kargil town. He explained that this shortage is primarily due to the increasing population and rural-to-urban migration. The issue has been attributed to rising water demand, limited water resources, and infrastructure challenges. This ongoing problem significantly impacts the daily lives of urban residents by limiting their access to reliable and clean water, with profound implications for public health, sanitation, and overall quality of life. There are three types of water tankers used to supply drinking water: mini water tankers, medium water tankers, and large water tankers. The mini tanker has a capacity of 6,000 L, the medium water tanker has a capacity of 10,000 L, and the large water tanker has a capacity of 12,000 L. On a daily and alternative basis, the PHE department supplies drinking water using 19 water tankers across all 13 wards and the newly developed settlements of Kargil town (Fig. 6.2).

Water tankers supply water to urban households in the early morning. However, tankers-based water supply becomes problematic in Wards 1, 2, 4, 8, 10, 12, and 13 due to office hours, traffic congestion, and increased market activities during the day. In contrast, Wards 3, 5, 6, 7, and 11 generally do not require water tanker

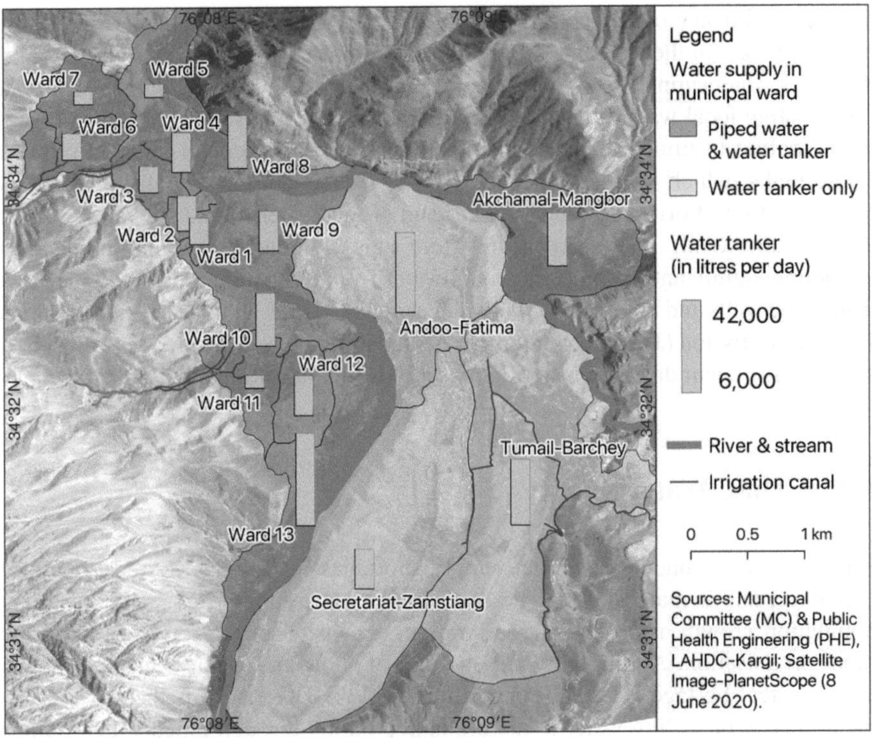

Fig. 6.2 Urban water supply across municipal wards in Kargil town, Ladakh. *Draft and cartography* Altaf Hussain

supply during the summer if there has been heavy snowfall in the winter. Simultaneously, the Kurbathang Plateau has experienced significant growth in residential colonies, as well as administrative and commercial buildings (Hussain et al., 2023). These buildings and newly developed infrastructure have placed immense pressure on the already strained water supply system. The resulting rise in water demand has further intensified pressure on the limited resources available in this town. For instance, the newly established hotels and guesthouses are using borewells to extract the groundwater, which will have dire repercussions on the hydrological system. Figure 6.3 shows how urban households access water using water storage tanks and hand pumps. Moreover, according to an interview conducted with a local migrant (Iftikhar Hussain, on 8 August 2022), the water supply through water tankers does not fulfil the daily water requirements of urban households. Each household is provided with either 200 or 500 L of water on a daily or alternate basis, depending on the locality's requirements. A household must pay ₹600 for 200 L of water for six months, which becomes ₹1,200 in a year. If a household requires 500 L of water, it must pay ₹1,500 for six months, which becomes ₹3,000 annually.

Ensuring a safe and adequate supply of drinking water is crucial for the socio-economic well-being of communities. Access to clean water is essential not only

Fig. 6.3 Water storage and access in the municipal areas of Kargil town. **e** Water storage tanks along the roadside at Balti bazaar. **f** Using a hand pump to obtain water from a borewell at Thaskangrong bazaar. **g** Water storage tanks in the newly constructed colony of the Kurbathang Plateau. **h** Water storage tanks at the road junction in Baroo bazaar. Photographs taken by Altaf Hussain on 23 August 2021

for health and hygiene but also for livelihoods, tourism, and overall quality of life. In Kargil town, however, the existing water supply system has failed to meet both quality and quantity standards, creating significant challenges for the urban population. Key challenges such as limited surface water availability, pipeline freezing during winter, reliance on water tankers, and inadequate storage infrastructure have exacerbated the water crisis. The town's high-altitude location and harsh climatic conditions intensify these problems, making water supply management even more difficult. Additionally, the rugged terrain, seasonal fluctuations in water sources, and the consequences of climate change further stress the water supply system. The retreat of glaciers and shifting precipitation patterns have resulted in unpredictable water availability, complicating long-term planning and resource management. As a result, many urban residents are forced to rely on alternative and often unreliable water sources, which disrupt daily life. The lack of efficient water management systems and limited technological interventions in this remote area make the issue even more difficult to address. Tackling these water supply challenges is crucial for ensuring sustainable urban development in Kargil town.

6.3 Municipal Solid Waste Management

Municipal solid waste management in Kargil is carried out daily under the supervision of the MC of LAHDC-Kargil. An interview was conducted on 5 September 2021 with official members of the MC in Kargil, Ladakh. According to Executive Officer (Shafaqat Ali Khan) and Junior Engineer (Mohammad Abass), municipal workers are assigned to all wards for waste collection and street cleaning. These municipal workers primarily clean roads, streets, and public spaces by sweeping and gathering waste into bins or designated collection points. They also collect waste from these points, public bins, and residential areas. The cleaning and waste collection activities are carried out in the early morning hours, typically between 6:00 a.m. and 9:00 a.m., to ensure that streets are cleaned prior to the start of daily routines.

The MC collects municipal solid waste from urban households in all wards. Most solid waste is generated from residential, commercial, and institutional areas. However, domestic animals consume compostable waste at the household level, depending on the animals being reared, which reduces the total volume of compostable material that needs to be collected. There has been an increasing amount of solid waste due to modernisation and changing lifestyles. The solid waste collected becomes difficult to segregate due to the mixing of dry and wet wastes at the source. Once collected, the solid waste is loaded and transported to the waste disposal site using municipal waste collection vehicles. The fleet consists of one refuse compactor, five dumper trucks, and several tippers for collecting solid waste. The town is approximately 12 km away from the Kurbathang waste disposal site. At the waste disposal site, rag pickers segregate the solid waste into dry and wet categories. Wet waste is dumped into a bottomless dig pit, which is 3 m deep, and covered with soil through a landfilling process. Dry waste (cardboard, plastics, etc.) and other recyclables are sent to Srinagar, Jammu, and other locations for recycling. A solid waste management plant is planned for construction at the Kurbathang waste disposal site to treat and manage solid waste more efficiently. The plant is expected to improve waste segregation, recycling, and treatment, thereby contributing to a more sustainable solid waste management system. An illustration of solid waste management challenges in Kargil town is provided in Fig. 6.4.

The MC Kargil of LAHDC-Kargil is responsible for overseeing a solid waste management system, which includes the efficient collection and transportation of waste, proper treatment and segregation, promotion of recycling initiatives, and the development of sanitary landfills for safe disposal. These tasks are carried out by municipal workers under its supervision. Despite ongoing efforts, open waste remains a significant issue in Wards 1, 2, 4, 8, 10, and 12, primarily due to the concentration of commercial zones in these areas. Additionally, waste accumulation is noticeably higher in newly developed colonies, such as Poyen Colony, Baroo Colony, Haidery Mohalla, and Andoo Colony, compared to the older, established parts of the town. A major challenge contributing to this issue is the limited sense of civic responsibility among the public. Many residents perceive solid waste management as solely the responsibility of MC Kargil, rather than a shared duty that requires active community

Fig. 6.4 Solid waste management issues in Kargil town. **i** Waste bin of MC on the street near Hussaini Park. **j** Dumping of solid waste on the roadside of Changchik bazaar. **k** Garbage trucks of MC Kargil near Fatima Chowk. **l** Disposal of solid waste in deep drains. **m** Open burning of solid waste. **n** Landfilling of solid wastes at Kurbathang disposal site. Photographs taken by Altaf Hussain between July and September 2022

participation. This lack of awareness and engagement exacerbates waste management challenges, particularly in newly developed residential areas that often lack proper waste collection services.

Furthermore, India generates approximately 62 million tonnes of municipal solid waste each year, posing substantial environmental challenges (Ghosh, 2016). In Ladakh, municipal waste has increasingly become an urban challenge, especially in the towns of Kargil and Leh. Solid waste is generated primarily from urban households, commercial buildings, hotels and guesthouses, institutions, shops, and restaurants. The total waste generation was estimated to be approximately 15.66 tonnes per day during summer and around 12.37 tonnes per day during winter (Kargil Municipal Committee, 2020). On average, about 6.43 tonnes of waste is collected daily and transported to the existing dumpsite at Kurbathang Plateau in Kargil town (Kargil Municipal Committee, 2020). Waste generation is expected to rise due to population growth and increasing per capita waste production. It is projected to grow from 25 tonnes per day to 63.43 tonnes per day over the next 20 years (Kargil Municipal Committee, 2020). There are currently no proper arrangements in place to manage hazardous effluents generated by hospitals, households, commercial activities, and small-scale industries, which have severe environmental impacts. Leh town has witnessed a massive influx of tourists, resulting in a large amount of waste being generated and intensifying environmental pollution (Dar et al., 2019). The increasing production of municipal waste in Ladakh, fuelled by urbanisation, tourism, and population growth, presents serious environmental concerns. In the absence of adequate waste management systems, environmental deterioration in Leh and Kargil is expected to escalate. Immediate implementation of effective waste management strategies is essential to addressing these issues and promoting sustainable urban development in Ladakh's towns.

6.4 Urban Deep Drains

The urban households do not have proper deep drains, except for a few located along the roadsides of the main bazaar in Kargil town. Several deep drains have been constructed and upgraded under the district plan through the IDMT scheme. The following deep drains have been developed as part of this initiative (Fig. 6.5): (o) construction of a deep drain from Khomeini Chowk to Hussaini Park; (p) upgradation of the deep drain from Lal Chowk to Poyen bridge; (q) construction of a deep drain from Choposa to Poyen bazaar; (r) construction of a deep drain from Government Boys Higher Secondary School to Sher-bagh; (s) construction of a deep drain from the circular road junction to bypass Shilikchey bridge; (t) the existence condition of the newly developed deep drain at the new taxi and bus stand. These deep drains were implemented in the town of Kargil by the R&B wing of PWD, LAHDC-Kargil, with an investment of more than ₹4 crore in Ladakh. This investment aimed at improving the town's drainage infrastructure and addressing the challenges of urban sanitation. Additionally, efforts are being made to expand sewer and septage management systems to develop an integrated urban drainage network for the town. However, implementing effective drainage systems remains challenging due to the difficult terrain, existing infrastructure constraints, and erratic weather patterns exacerbated by climate change.

The existing deep drains are not designed to handle greywater effluent from households. They are solely meant to collect stormwater from the roads. There is no proper drainage system in place in any of the municipal wards of Kargil town. The town has a network of roadside deep drains constructed by LAHDC-Kargil, which are maintained by the MC. These deep drains are primarily located along roads and pathways in the bazaar area. However, blockages and sewage overflows have been observed due to improper maintenance and a lack of public awareness. These deep drains

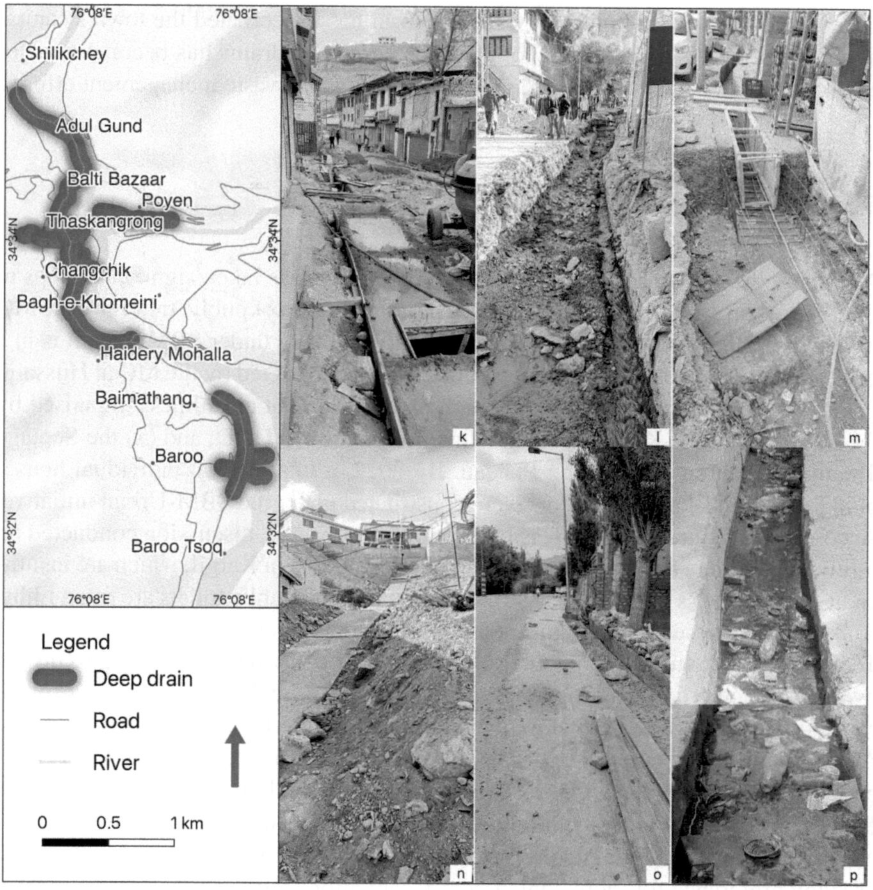

Fig. 6.5 Construction and upgradation of deep drains in the municipal areas of Kargil town. Photographs taken by Altaf Hussain between June and August 2021

often get clogged with garbage and waste disposed of by people. As a result, the drains, along with the accumulated waste, discharge into the Suru River, leading to pollution. This contamination not only degrades the river's water quality but also poses significant environmental and health risks to the local population. The accumulation of waste further blocks the flow of water, creating unpleasant odours and contributing to the deterioration of the local environment. Additionally, hotel and restaurant operators, as well as shopkeepers, are commonly seen using deep drains for waste disposal and sewage purposes. Originally designed to manage stormwater runoff and wastewater, these drains have increasingly become dumping grounds for solid and liquid waste. In addition, households and migrant tenants, many of who live in rented accommodations, contribute to the pollution of these deep drains by disposing of household waste, sewage, and other materials directly into them. The widespread practice of improper waste disposal has exacerbated the town's sanitation conditions. As a result, the pollution of these deep drains has become a major environmental concern, further complicating the town's waste management efforts.

6.5 Public Toilets

As part of the SBM-Urban initiative, LAHDC-Kargil has taken significant steps to improve urban sanitation infrastructure. The construction of public toilets by the MC is shown in Fig. 6.6: (u) a public toilet built at Daythang under SBM, which is in a poor and unhygienic condition; (v) a public toilet constructed by the MC at Hussaini Park; (w) an awareness-raising campaign on environment cleanliness organised by the MC of LAHDC-Kargil under the direction of the H&UDD; and (x) the Septage Treatment Plant at Kurbathang Plateau. By 2022, a total of 400 individual household latrines and 12 public toilets were constructed under the SBM-Urban initiative. According to Fayaz Ali (a professional architect of Kargil, discussion conducted on 6 July 2022), there are only a few public toilets available in Kargil, which are insufficient for the town's population. The existing municipal public toilets are often filthy, stinky, and unhygienic, highlighting the need for regular cleaning and maintenance. To address this issue, more public toilets should be constructed, adhering to proper bylaws, particularly in parking areas, recreational sites, and shopping complexes, to cater to the needs of the growing population and ensure better sanitation facilities. According to Mohd Ismail, a local shopkeeper and owner of a small plywood enterprise (discussion conducted on 17 September 2021), modern toilets are becoming a trend in newly constructed buildings. Dry compost toilets, which were once used as manure (lut) in agricultural fields, are rarely seen in town today. This shift reflects the growing preference for more contemporary sanitation solutions, though it may also signal the decline of traditional practices that were once integral to local agricultural systems.

Due to modernisation, the traditional toilet system is being replaced by the modern flush toilet system, particularly in urban centres. These flush toilets are commonly installed in newly constructed residential, administrative, and commercial buildings. However, the widespread use of water-based flush systems has led to increased moisture in buildings located on sloped areas, making them more vulnerable to landslides. In the residential buildings of Silmo Colony, Andoo Colony, and Haidery Mohalla, which are already vulnerable to landslides, the use of water-based modern flush toilet systems has resulted in the wetting of barren soil on the premises, further increasing the susceptibility of these settlements to landslides. For managing liquid wastes, a Faecal Sludge Treatment Plant (FSTP) has been constructed by the PHE department of LAHDC-Kargil. The FSTP is designed to handle and treat faecal sludge generated from septic tanks across the town. To facilitate the collection of this waste, the municipality has deployed four mini trucks equipped with suction pumps. When the septic tanks of town residents become full, they contact the MC, which then dispatches one of these mini trucks to pump out the waste from the

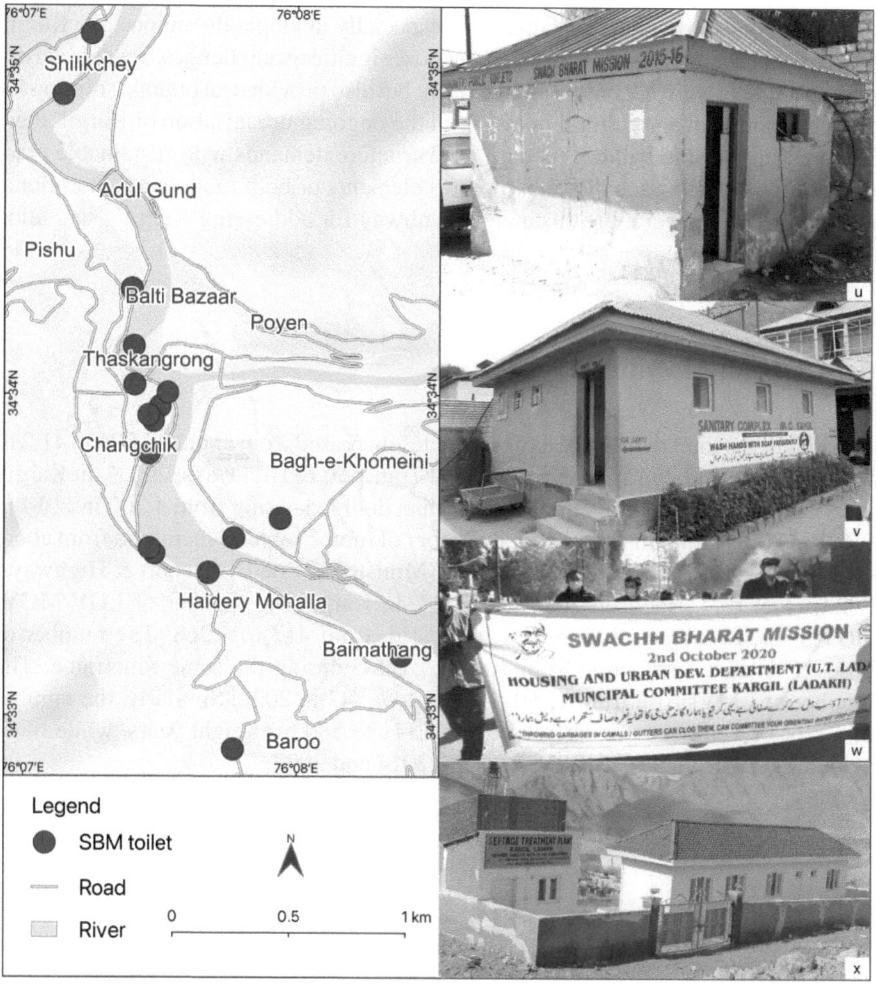

Fig. 6.6 Spatial distribution of public toilets constructed under SBM in Kargil town. Photographs taken by Altaf Hussain in August 2021 (**u**, **v**) and September 2022 (**x**); photograph (**w**) obtained from the MC office in 2022

septic tanks. The waste is then transported and safely disposed of at the FSTP for proper treatment and processing. The MC charges the residents a fee for this waste management service, which helps in maintaining the sanitation infrastructure. With the implementation of the AMRUT 2.0 scheme, a sewage treatment plant will be established in the coming years, which will enhance the town's waste management infrastructure and help meet the growing demand for wastewater treatment.

Moreover, the adoption of modern toilets can be attributed to various factors, including the influence of government initiatives like the SBM, increased awareness about hygiene, and aspirations for urban lifestyles. However, this transition also raises questions about sustainability, especially in mountain regions like Kargil, where water availability and management pose significant challenges. Traditional dry compost toilets were not only water-efficient but also provided an organic solution for soil enrichment in agricultural fields. With the ongoing urbanisation of Kargil, there is a pressing need to balance modern infrastructure demands with sustainable practices. A hybrid approach that incorporates elements of both modern and traditional systems could offer a more sustainable pathway for addressing Kargil's sanitation and agricultural needs.

6.6 Urban Mobility

In Ladakh, the total number of motor vehicles increased from about 27,096 to 41,216 between 2018 and 2022 (Ministry of Road Transport & Highways, 2023). In Kargil, the total number of motor vehicles more than doubled, rising from 4,300 in 2014 to 9,401 in 2022, while In Leh, the total number of motor vehicles increased from about 20,639 to 31,815 between 2018 and 2022 (Ministry of Road Transport & Highways, 2023) (Fig. 6.7). Between 2014 and 2022, in Kargil, the number of LGV/MGV/HGV vehicles increased more than threefold, from 417 to 1,268. The number of buses/taxis/maxi cabs grew from 1,373 to 1,935 during the same timeframe. The number of LMVs rose from 1,691 in 2014 to 4,547 in 2022. Similarly, the number of motor bicycles/scooters increased from 341 to 586 over eight years, while other vehicles grew from 478 to 1,065 between 2014 and 2022.

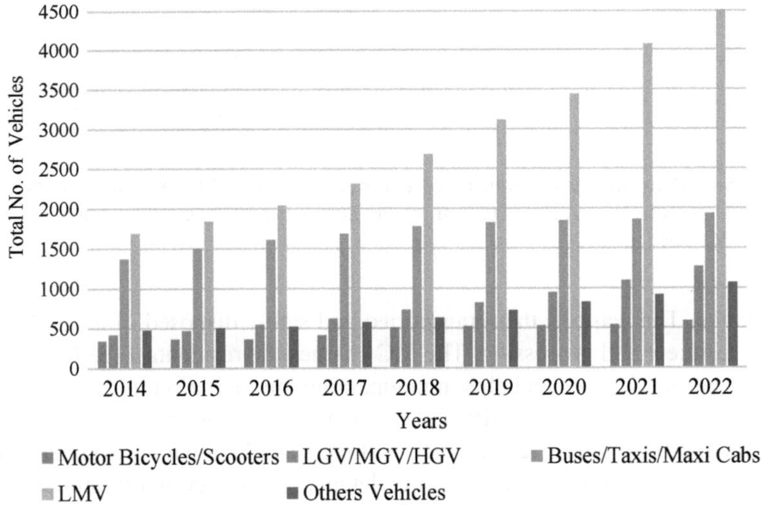

Fig. 6.7 Total number of registered motor vehicles in Kargil from 2014 to 2022. *Sources* Regional Transport Office Kargil of LAHDC-Kargil and MoRTHs, India; Light Motor Vehicles (LMV), Light Goods Vehicles (LGV), Medium Goods Vehicles (MGV), and Heavy Goods Vehicles (HGV)

There are a total of eight parking stands in Kargil town for buses, vans, cars, taxis, etc. The parking stands are located at Kabadi Nallah, Old Taxi Stand, Bakpalongma, Baroo Market, Conference Hall Baroo, Qatilgah Complex, Balti Bazaar (Drass Taxi Stand), and Hussaini Park in Kargil town. However, these parking stands have less capacity in comparison to the number of increased vehicles. They can accommodate no more than 600 vehicles. This has led to heavy traffic congestion, compounded by unauthorised parking on the streets and roads of Kargil town.

In addition, the roads are very narrow and haphazard parking by various types of vehicles is also common in some urban areas. Such haphazard parking and roadside encroachments are particularly prevalent in the bazaars (Changchik, Thaskangrong, Changrah, Poyen, and Baroo) of Kargil town, significantly disrupting road capacity. Vehicle numbers rise during the summer months due to increased tourist activity, the seasonal opening of the Srinagar-Kargil-Leh highway, and enhanced trade interactions. Traffic blockages and congestion issues are common, particularly during the morning and afternoon peak hours. This is mainly due to poor road design by local engineers, who lacked the necessary expertise to develop an adequate road network system. The narrow roads also make it difficult for two-way traffic to pass smoothly. There is a need for wider and better-constructed roads. In some residential areas, houses are only accessible by narrow alleyways, which makes it extremely difficult for ambulances and fire engines to navigate during emergencies. Additionally, there is a lack of a planned and properly organised public transport system. The increasing number of motor vehicles, particularly private cars on narrow streets and roads, is creating bottlenecks and hindering the smooth flow of traffic (Fig. 6.8).

The development of road infrastructure has led to an increase in the arrival of domestic and international tourists in Ladakh. The tourism sector is rapidly expanding in Kargil due to the growth of tourism activities, commercial interactions, and improved road connectivity. Tourist arrivals in Kargil have surged dramatically from 4,325 in 1976 to 225,543 in 2022, resulting in an increase in hotels, guest houses, and travel agencies (Hussain et al., 2023). Heavy automobiles are now commonly seen, particularly in the capital towns of Ladakh, reflecting the growing urbanisation and the increasing demand for transportation infrastructure. Due to the rise in the number of automobiles, traffic jams on urban road networks, especially in the bazaars of both capital towns, have become more frequent. The main causes of automobile congestion on public roads include a lack of proper parking spaces, inadequate traffic control devices, the absence of traffic rules and regulations, and limited investment in traffic infrastructure.

For the beautification of the town, footpaths/sidewalks were constructed with well-finished tiling in all wards of Kargil. Under IDMT scheme, 150 footpaths were developed, along with 910 streetlights installed in the municipal wards between 2020 and 2022. An estimated amount of ₹2.08 crore was spent by the R&B department of LAHDC-Kargil. The footpaths were constructed with well-finished pavement tiles in residential areas like Poyen, Bagh-e-Khomeini, Changchik, Thaskangrong, Lankore, Goma Kargil, Balti Bazar, and Baroo (Fig. 6.9). These sidewalks/footpaths were primarily developed along commercial streets, residential streets, and market (bazaar) roads. In Kargil town, however, urban footpaths/sidewalks are poorly developed due

Fig. 6.8 Traffic and parking challenges in Kargil town. Photographs taken by Altaf Hussain between July and September 2022

to the challenging terrain and the scarcity of flat land for infrastructure development. The lack of adequate space for pedestrian pathways is further exacerbated by steep slopes, narrow roads, and limited urban planning resources. As a result, pedestrians face difficulties moving safely throughout the town, particularly in the marketplaces. To address these challenges, the town requires innovative approaches to incorporate pedestrian pathways that preserve its natural landscape while also addressing the increasing demand for urban space.

According to the Assistant Regional Transport Officer of LAHDC-Kargil, Khadim Hussain (interview conducted on 18 August 2021), pedestrian spaces in Kargil are highly constrained. The limited footpath areas are further reduced by encroachments, with vendors, shopkeepers, business owners, and private vehicle owners occupying significant portions of these spaces. As a result, pedestrians are often unable to use footpaths and are forced to walk on the roads, which disrupts the flow of vehicular traffic in the town. This encroachment not only causes congestion but also compromises the safety and convenience of pedestrians, highlighting a major urban development challenge in Kargil. The congestion of vehicular traffic in the central town is particularly evident from Iqbal bridge (zamba) to the bypass road during school and office hours. In addition, electric poles, wires, water pipes, and water tanks and other elements of encroachment are commonly observed on footpaths and roadsides in the town. The informal and unregulated use of public spaces significantly hampers efforts to establish planned pedestrian and ensure efficient urban mobility. Footpaths/

Fig. 6.9 Urban beautification in Kargil town. **y** Development of Gandhi Park by laying pavement tiles at Baroo. **z** Laying of tiles on the footpath/sidewalk along the roadside at Changchik bazaar. Photographs taken by Altaf Hussain in September 2022

sidewalks need to be designed according to the guidelines set by the Indian government, and robust regulations must be implemented to create a proper pedestrian zone or system.

Moreover, urban infrastructure in Kargil has been under considerable pressure due to population growth and rural-to-urban migration. Being the main centre of economic growth, unplanned development has resulted in an increase in solid waste, lack of water supply, traffic congestion, toilets problems, poor drainage, and inadequate wastewater treatment. In addition, there are no proper recreational or public space facilities. In Leh, there is a lack of critical urban infrastructure like transportation, water supply, housing, solid waste management, sanitation, and the provision of recreational facilities. The traditional character of Leh town is in decline due to unplanned development and gross negligence of the environment (LAHDC-Leh, 2005). However, Leh remains of central importance due to its crucial infrastructure, such as transportation, healthcare, education, commerce, and administrative facilities (Müller & Dame, 2016). There is a need to provide high-quality, modern infrastructure with the efficient delivery of public services to the people of Ladakh. Additionally, a robust approach to inclusive and sustainable development is needed. The UT administration of Ladakh and LAHDCs must ensure and adopt sustainable development practices to make the region's towns safe, inclusive, resilient, and sustainable.

References

All India Radio. (2022, May 10). *Centre sanctions Rs 120 crore for Ladakh under AMRUT 2.0.* State News. https://newsonair.gov.in/News?title=Centre-sanctions-Rs-120-crore-for-Ladakh-under-AMRUT-2.0&id=440674

Ballaney, S. (2008). *The town planning mechanism in Gujarat, India.* The World Bank. https://hcp-com.b-cdn.net/wp-content/uploads/2021/06/Town-Planning-of-Gujarat_Research-Paper.pdf

Dar, S. N., Wani, M. A., Shah, S. A., & Skinder, S. (2019). Identification of suitable landfill site based on GIS in Leh, Ladakh region. *GeoJournal, 84*(6), 1499–1513. https://doi.org/10.1007/s10708-018-9933-9

Ghosh, S. K. (2016). Swachhaa Bharat mission (SBM)—A paradigm shift in waste management and cleanliness in India. *Procedia Environmental Sciences, 35*, 15–27. https://doi.org/10.1016/j.proenv.2016.07.002

Gondhalekar, D., & Akhtar, A. (2015). Down the drain or back to the roots? Political ecology of the water-energy-food nexus visualized using GIS in Leh XE town, Ladakh, India. In K. Goh & E. Chu (Eds.), *Projections 11—Global ecologies: Politics, planning, and design* (pp. 77–96). The MIT Journal of Planning.

Gondhalekar, D., Nussbaum, S., Akhtar, A., & Kebschull, J. (2015). Planning under uncertainty: Climate change, water scarcity and health issues in Leh town, Ladakh, India. In W. Leal Filho, & V. Sümer (Eds.), *Sustainable water use and management* (pp. 293–312). Springer International Publishing. https://doi.org/10.1007/978-3-319-12394-3_16

Housing & Urban Development Department. (2012, 2013). *Annual state plan.* Government of Jammu & Kashmir. http://jkhudd.gov.in/pdfs/Annual%20Plan%202012-13%20H&UDD%20-%20write%20up.pdf

Hussain, A., Schmidt, S., & Nüsser, M. (2023). Dynamics of mountain urbanisation: Evidence from the Trans-Himalayan town of Kargil, Ladakh, India. *Land, 12*(4), 920. https://doi.org/10.3390/land12040920

IDSMT. (1995). *Integrated development of small and medium towns (IDSMT).* Ministry of Urban Affairs & Employment, Department of Urban Development. https://mohua.gov.in/upload/uploadfiles/files/90.pdf

Jammu and Kashmir State. (2017). *State annual action plan (SAAP): Atal mission for rejuvenation and urban transformation (AMRUT).* http://164.100.87.10/writereaddata/saap/Saap17_20/JammuANDKashmir.pdf

Kargil Development Authority. (2007). *Prime Minister Reconstruction Programme (PMRP).* Ladakh Autonomous Hill Development Council Kargil, Jammu & Kashmir Government.

Kargil Development Authority. (2011). *Development of tourist infrastructure in Kargil district under destination development scheme (phase-I).* Ladakh Autonomous Hill Development Council Kargil, Jammu & Kashmir Government.

Kargil Municipal Committee. (2020). *Draft solid waste management (SWM) detailed project report (DPR) for Kargil under Swachh Bharat Mission.* Urban Local Bodies, The Adminstration of Union Territory of Ladakh.

LAHDC-Leh. (2005). *Ladakh 2025—A road map for progress and prosperity.* Ladakh Autonomous Hill Development Council, Leh. https://cdn.s3waas.gov.in/s3291597a100aadd814d197af4f4bab3a7/uploads/2018/06/2018061732.pdf

Lahmouri, M., Drewes, J. E., & Gondhalekar, D. (2019). Analysis of greenhouse gas emissions in centralized and decentralized water reclamation with resource recovery strategies in Leh town, Ladakh, India, and potential for their reduction in context of the water–energy–food nexus. *Water, 11*(5), 906. https://doi.org/10.3390/w11050906

Ministry of Housing & Urban Poverty Alleviation. (2015). *Pradhan Mantri Awas Yojana (housing for all-urban): Scheme guidelines.* Government of India. https://pmaymis.gov.in/PDF/HFA_Guidelines/hfa_Guidelines.pdf

Ministry of Road Transport & Highways. (2023). *Parivahan Sewa.* Vahan Dashboard. https://vahan.parivahan.gov.in/vahan4dashboard/

Ministry of Urban Development. (2015). *Atal mission for rejuvenation and urban transformation (AMRUT): Mission statement & guidelines*. Government of India. http://www.amrut.gov.in/upload/uploadfiles/files/AMRUT%20Guidelines%20(1).pdf

Müller, J., & Dame, J. (2016). Small town, great expectations: Urbanization and beautification in Leh. *South Asia Multidisciplinary Academic Journal, 14*. https://doi.org/10.4000/samaj.4208

Müller, J., Dame, J., & Nüsser, M. (2020). Urban mountain waterscapes: The transformation of hydro-social relations in the trans-Himalayan town Leh, Ladakh, India. *Water, 12*(6), 1698. https://doi.org/10.3390/w12061698

National Commission on Urbanisation. (1988). *Report of the national commission on urbanisation, 1988, (Vol. I–VII)*. Government of India.

NITI Aayog. (2021). *Reforms in urban planning capacity in India*. Governement of India. https://www.niti.gov.in/sites/default/files/2021-09/UrbanPlanningCapacity-in-India-16092021.pdf

Spodek, H. (2013). City planning in India under British Rule. *Economic and Political Weekly, 48*(4), 53–61. https://www.jstor.org/stable/23391350

The Constitution (Seventy-Fourth Amendment) Act, 1992. (1992). https://mohua.gov.in/upload/uploadfiles/files/ConsAct74.pdf

World Bank. (2013). *Urbanization beyond municipal boundaries: Nurturing metropolitan economies and connecting peri-urban areas in India*. The World Bank. https://doi.org/10.1596/978-0-8213-9840-1

The Ladakh Autonomous Hill Development Councils Act, 1997. (1997). https://faolex.fao.org/docs/pdf/IND227527.pdf

Chapter 7
Dynamics of Land Cover Transformation

Abstract This chapter explores the transformation of LULC in the high mountain region of Ladakh. The primary catalyst for this transformation has been the development of road connectivity. The opening of motorised traffic in 1960 and the introduction of international tourism in 1974 have significantly contributed to the socio-economic changes in Kargil, Ladakh. The study focuses on Kargil town and the urbanised settlements of Padum, Sankoo, and Drass to illustrate the changes in LULC since the 1960s. These areas have experienced rapid urbanisation driven by improved road connections, a growing tourism sector, and infrastructure development. These developments have enhanced connectivity and accessibility, thereby promoting economic activities and increasing external intervention. These changes, in turn, have led to significant shifts in land use and agricultural patterns, reflecting the dynamic transformation of the region's socio-economic landscape. There has been a notable shift from agricultural to non-agricultural activities, with many cultivators transitioning to more economically driven pursuits.

Keywords Built-up area · Barren area · Irrigated land and vegetation · LULC · Transformation · Occupational distribution · Connectivity · Tourism · Kargil · Padum · Sankoo · Drass · Ladakh

7.1 Key Development Interventions in Ladakh

Since the opening of Ladakh to motorised traffic in 1960 and international tourism in 1974, rapid urban transformation has been observed across the towns of Leh and Kargil, as well as in their surrounding villages. Road connectivity to Ladakh played a significant role in its development and integration with the rest of India. It served as the primary means of accessibility and intervention in this remote region. The improved road connection facilitated the movement of goods, people, and services, including trade and tourism, and integrated the region into the national economy. This resulted in a large-scale inflow of defence personnel, impacting the local economy throughout Ladakh, with increased demand for horses, wage labour, and the arrival of industrial

A. Hussain, *Urbanisation and Development Challenges in Kargil, Ladakh*, Advances in Asian Human-Environmental Research, https://doi.org/10.1007/978-3-032-05667-2_7

commodities (Crook & Osmaston, 1994). Hiring local wage labour for construction and services, along with the growing demand for locally produced goods, acted as a major catalyst for change. This transformation introduced various new items, including electricity, automobiles, Western clothing, coffee, tea (cha), sugar (khara), rice (bras), jerrycans, and lanterns. The most significant change, however, was the influx of cash capital and its distribution throughout the region. This enabled many local people to purchase industrial commodities and led to a major shift in consumption patterns (Crook & Osmaston, 1994).

Additionally, the year 1974 marked a turning point in Ladakh's history, as the Indian government decided to open most of the border areas to tourism. Before that, only Indian nationals were allowed to enter Ladakh with a clearance permit. The opening of Ladakh to tourism significantly influenced the region's socio-economic and cultural dynamics. The influx of tourists brought new economic opportunities, such as increased demand for hospitality services and created employment opportunities for locals. However, it also resulted in challenges concerning the preservation of cultural heritage and the region's environmental sustainability. The significant growth of tourism and the development of road connectivity (NH-1,[1] NH-301, and NH-3) have greatly affected the settlements along these roads, leading to increased economic activities and changes in local livelihoods (Fig. 7.1). The improved road network has strengthened Ladakh's connectivity, facilitating a greater exchange of goods, information, and movement of people. This has also impacted social, economic, and cultural life. Improved accessibility and strategic importance have accelerated the development of infrastructure. Additionally, the central government and LAHDCs have also invested in several road infrastructure projects, particularly the upgradation of district roads and national highways.

This chapter investigates Kargil town and the urbanised settlements of Padum, Sankoo, and Drass, located in the Kargil district of Ladakh, all of which have undergone rapid urbanisation. Road connectivity and tourism have been the primary drivers of socio-economic transformation in the lives of the people. The region's geopolitical importance as a borderland has led to the rapid construction of roads in remote areas, fluctuations in the tourism sector, changes in political governance, and significant infrastructure development. Recently, a shift from agricultural to non-agricultural activities has been observed, with many cultivators transitioning to economically driven pursuits. The opening of the Srinagar-Kargil-Leh road in 1960 facilitated the influx of subsidised food commodities into Ladakh, leading to increased dependence on these external supplies. This shift contributed to the decline of subsistence agriculture, which had once been the sole means of livelihood survival. As a result, traditional agricultural practices diminished, and local communities began to rely more on external sources for their basic needs, altering the region's economic and cultural landscape.

[1] NH-1 runs between the UTs of Jammu & Kashmir and Ladakh in India. In 2010, the old name of NH-1D was renumbered as NH-1 through new numbering system of National Highways by MoRTH in India.

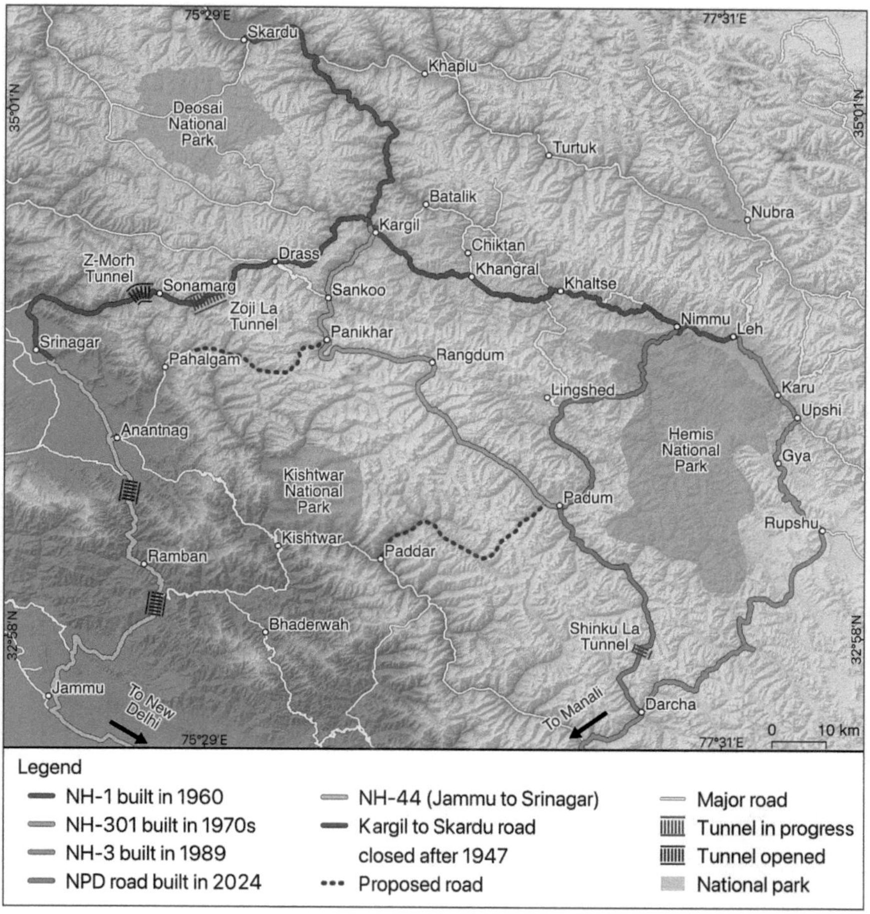

Fig. 7.1 Development of road connectivity in Ladakh. *Sources* SRTM (11 February 2000); OSM. *Draft and cartography* Altaf Hussain

In the past, settlements were primarily situated in areas with easy access to water and fertile land, often along river valleys. These locations provided the resources necessary for agriculture, trade, and protection from the harsh mountain environment. Over time, these settlements expanded slowly, relying heavily on traditional agricultural and trade practices, with limited infrastructure. In contrast, today's settlement patterns are shaped by factors such as road connectivity, tourism, and government policies. Many modern settlements have developed due to road networks, as seen in the cases of NH-1 and NH-301 (Fig. 7.2). Residential areas are primarily developing along roads, rivers, streams, and canals, as observed in all the settlements across Kargil district. Infrastructure improvements, such as schools, healthcare facilities, administrative buildings, and transportation, have also contributed to the growth of settlements across the region of Ladakh.

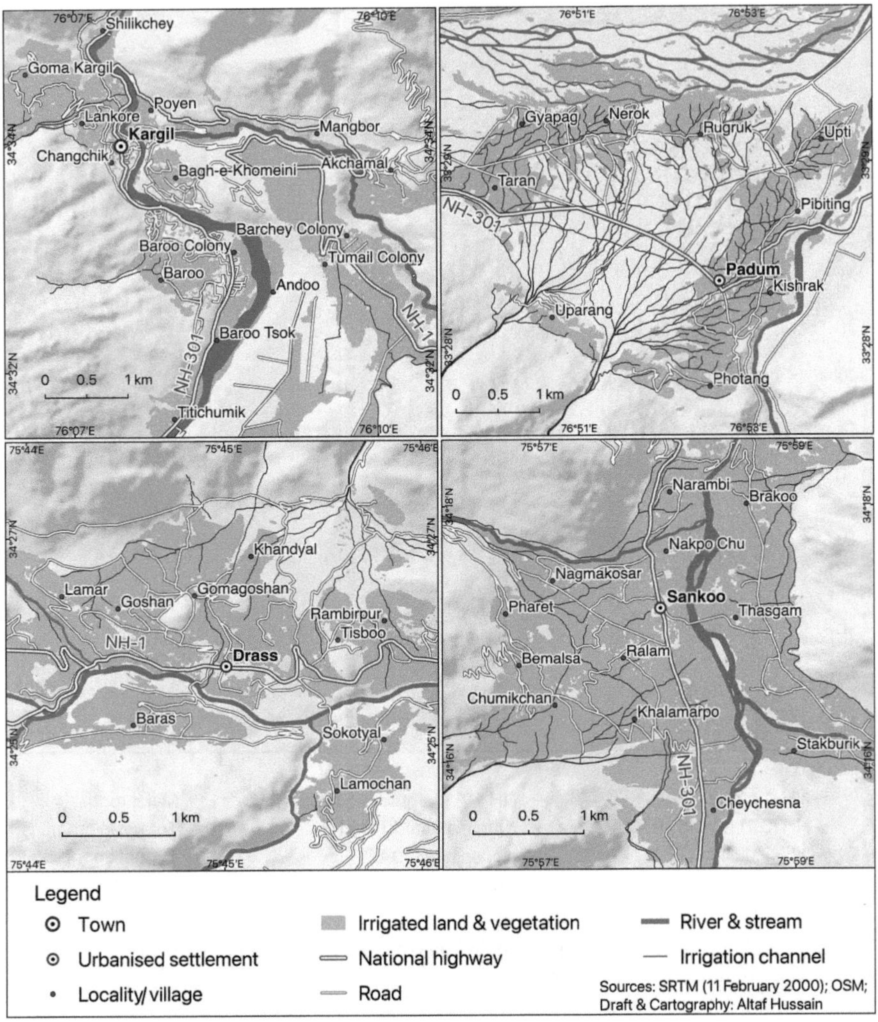

Fig. 7.2 Location of Kargil town and the urbanised settlements of Padum, Sankoo, and Drass in Ladakh

Settlements in the mountain region were traditionally formed in a clustered pattern but are now shifting to linear and dispersed forms. The low-lying mountain areas are the most suitable for the cultivation and settlement development. The formation of settlements is significantly influenced by factors such as climate, topography, water resources, and road accessibility. Barren land and sloped areas are typically preferred for construction activities, while irrigated land in valleys and terraces is ideal for agriculture. However, substantial building construction on irrigated land in Ladakh has reduced space for farming, impacting agricultural productivity and threatening local food security. This shift in land use is not only reducing the area available for crops but also disrupting traditional farming practices, thereby increasing reliance on imported food supplies and threatening the region's self-sufficiency.

In the Trans-Himalayan region, irrigation water is supplied through a gravity-flow system that harnesses runoff, meltwater from glaciers, and snowfields (Hill, 2014; Nüsser et al., 2012). Irrigation reservoir (rzing), water canal, and channel (yourba) play a significant role in irrigating cultivated land (jhing) and agricultural fields. In 1996, the WDP was implemented in Ladakh's arid, high-altitude terrain as a component of the Desert Development Programme (Mankelow, 2003). The initiative aimed to promote integrated rural development by enhancing irrigation infrastructure, introducing agricultural innovations, and supporting afforestation efforts (Nüsser et al., 2012). As a result of this programme, the government brought nearly 1,000 ha of land under irrigated cultivation in Ladakh's arid, high-altitude region (Rizvi, 2012). In addition to government agencies, several NGOs participated in the implementation of the WDP, undertaking development initiatives at the local level in collaboration with project implementing agencies (Nüsser et al., 2012).

In 2015–16, the LAHDC-Kargil introduced the PMKSY programme with the objective of strengthening the district's irrigation infrastructure. Under the District Irrigation Plan (DIP) of PMKSY, major crops such as barley (nas), wheat (kro), potato, peas, pulses, and vegetables have been irrigated, resulting in notable improvements in the socio-economic conditions of the local population. The implementation of PMKSY in Kargil district has resulted in an increase in the irrigated area for various crops, from 11,454 to 18,861 ha (LAHDC-Kargil, 2016). Under the DIP, 710 ha of land area were irrigated in the Zanskar block, 583 ha in the Sankoo block, and 1,380 ha from the existing irrigated area in the Drass block of Kargil district (LAHDC-Kargil, 2016). JJM, a central government initiative under the Ministry of Jal Shakti, was launched by Government of India in 2019. Under this scheme, assured water supply and irrigation water are provided to support the socio-economic growth of both rural and urban areas in Ladakh. Micro irrigation, lift irrigation, and drip irrigation schemes have been implemented to enhance field channels and canal irrigation infrastructure. The I&FC Division and PWD are the main implementing agencies under the LAHDCs in Ladakh. In the Kargil district, various snow check dams, water storage tanks, and water conservation structures are being constructed under a special development package by the I&FC Division of LAHDC-Kargil, funded by the UT administration of Ladakh.

7.1.1 LULC Change in Kargil

The transformation of the physical landscape began when Ladakh was opened to vehicular traffic in August 1960 (Zutshi et al., 1961). External intervention, both from governmental policies and outside forces, has led to profound changes in the physical and socio-economic landscape of the region over several decades. These interventions have caused significant changes in land cover, primarily driven by the rapid urbanisation and human activities. A drastic change in LULC can be observed between 1965 and 2020 in the Kargil town. The built-up area increased from 25 to 43 ha (1.28 ha per year) between 1965 and 1979, from 43 to 111 ha (3.24 ha per year) between 1979 and 2000, from 111 to 230 ha (5.95 ha per year) between 2000 and 2020, and from 25 to 230 ha (3.73 ha per year) over the entire period between 1965 and 2020 (Table 7.1). Consequently, the built-up area has grown more than ninefold over the past six decades. The irrigated land and vegetation increased from 451 to 488 ha (2.64 ha per year) between 1965 and 1979, from 488 to 818 ha (15.71 ha per year) between 1979 and 2000, from 818 to 856 ha (1.90 ha per year) between 2000 and 2020, and from 451 to 856 ha (7.36 ha per year) during the entire period between 1965 and 2020. The barren area decreased from 1488 to 1433 ha (3.99 ha per year) between 1965 and 1979, from 1433 to 1035 ha (18.95 ha per year) between 1979 and 2000, from 1035 to 878 ha (7.85 ha per year) between 2000 and 2020, and from 1488 to 878 ha (11.09 ha per year) during the entire period between 1965 and 2020. The Kurbathang Plateau played a key role in the expansion of irrigated land and vegetation cover. Previously barren and covered with only sparse, low-lying vegetation, the area was converted into a lush green plateau in the 1980s by channelling water from the Wakha River for irrigation (Khan, 2020; Rizvi, 2012).

Overall, Kargil town's built-up area expanded by 18 ha between 1965 and 1979, followed by an increase of 68 ha from 1979 to 2000. Between 2000 and 2020, an additional 119 ha was added. In total, the built-up area grew by 205 ha over the entire period from 1965 to 2020. Irrigated land and vegetation expanded by 37 ha between 1965 and 1979, and saw a substantial increase of 330 ha from 1979 to 2000, driven by the development of new irrigation canals. Following this major growth,

Table 7.1 LULC and its change detection from 1965 to 2020 in Kargil town

Area change (ha)	Years	Class		
		Built-up area	Irrigated land and vegetation	Barren area
Area	1965	25	451	1488
	1979	43	488	1433
	2000	111	818	1035
	2020	230	856	878
Area per year	1965–1979	1.28	2.64	3.99
	1979–2000	3.24	15.71	18.95
	2000–2020	5.95	1.90	7.85
	1965–2020	3.73	7.36	11.09
Change detection area	1965–1979	18	37	−55
	1979–2000	68	330	−398
	2000–2020	119	38	−157
	1965–2020	205	405	−610

Sources Satellite imagery: Corona (8 Oct 1965), Corona (17 June 1979), ASTER (4 September 2000), and PlanetScope (8 June 2020)

the irrigated land and vegetation remained largely stable, with a modest increase of 38 ha between 2000 and 2020. Nonetheless, residential and commercial structures have been built on areas that were previously irrigated land. In total, irrigated land and vegetation expanded by 405 ha between 1965 and 2020. Meanwhile, the barren area decreased by 55 ha between 1965 and 1979, and by 398 ha between 1979 and 2000. The significant decrease in barren area was mainly due to the irrigation system, which transformed barren land into irrigated land and vegetation between 1979 and 2000. A further decrease of 157 ha was observed in barren land from 2000 to 2020. In total, a reduction of 610 ha in barren area was recorded between 1965 and 2020. The spatial changes in LULC in Kargil town from 1965 to 2020 are clearly evident (Fig. 7.3), as reflected in the aerial view of the Baroo settlement (Fig. 7.4).

In Ladakh, irrigation is a fundamental requirement for practicing agriculture. Crops are cultivated through human-made water canals or channels (yourba), as water is abundant during the summer season in rivers and streams due to the melting of snow (kah) and glaciers. Most precipitation occurs in the form of snow (kah), which is critically important before the agricultural season. Climate elements such as temperature and precipitation play a significant role in agricultural practices, with a strong reliance on the irrigation system under the unique environmental conditions of Ladakh.

7.1.2 LULC Change in Padum

Padum is a small, urbanised settlement located on an alluvial fan, nearly 234 km away from the district capital. In Padum, the built-up area increased from 10 to 45 ha (1.40 ha per year) between 1980 and 2005, from 45 to 82 ha (2.31 ha per year) between 2005 and 2021, and from 10 to 82 ha (1.76 ha per year) over the entire period from 1980 to 2021. The irrigated land and vegetation increased from 742 to 869 ha (5.08 ha per year) between 1980 and 2005, from 869 to 990 ha (7.56 ha per year) between 2005 and 2021, and from 742 to 990 ha (6.04 ha per year) over the entire period from 1980 and 2021 (Table 7.2). Meanwhile, the barren area decreased from 5,157 to 4,995 ha (6.48 ha per year) between 1980 and 2005, from 4,995 to 4,837 ha (9.87 ha per year) between 2005 and 2021, and from 5,157 to 4,837 ha (7.80 ha per year) over the entire period between 1980 and 2021.

Padum experienced an increase in built-up area, with 35 ha detected from 1980 to 2005 and 37 ha detected from 2005 to 2021. Overall, a total of 72 ha of built-up area was detected between 1980 and 2021. Additionally, 127 ha of irrigated land and vegetation were observed from 1980 to 2005, and 121 ha from 2005 to 2021, resulting in a total increase of 248 ha in irrigated land and vegetation between 1980 and 2021. The barren area also decreased by 162 ha from 1980 to 2005 and by 158 ha from 2005 to 2021, leading to a total decrease of 320 ha in barren area between 1980 and 2021. The spatial transformation of LULC in Padum between 1980 and 2021 is evident (Fig. 7.5).

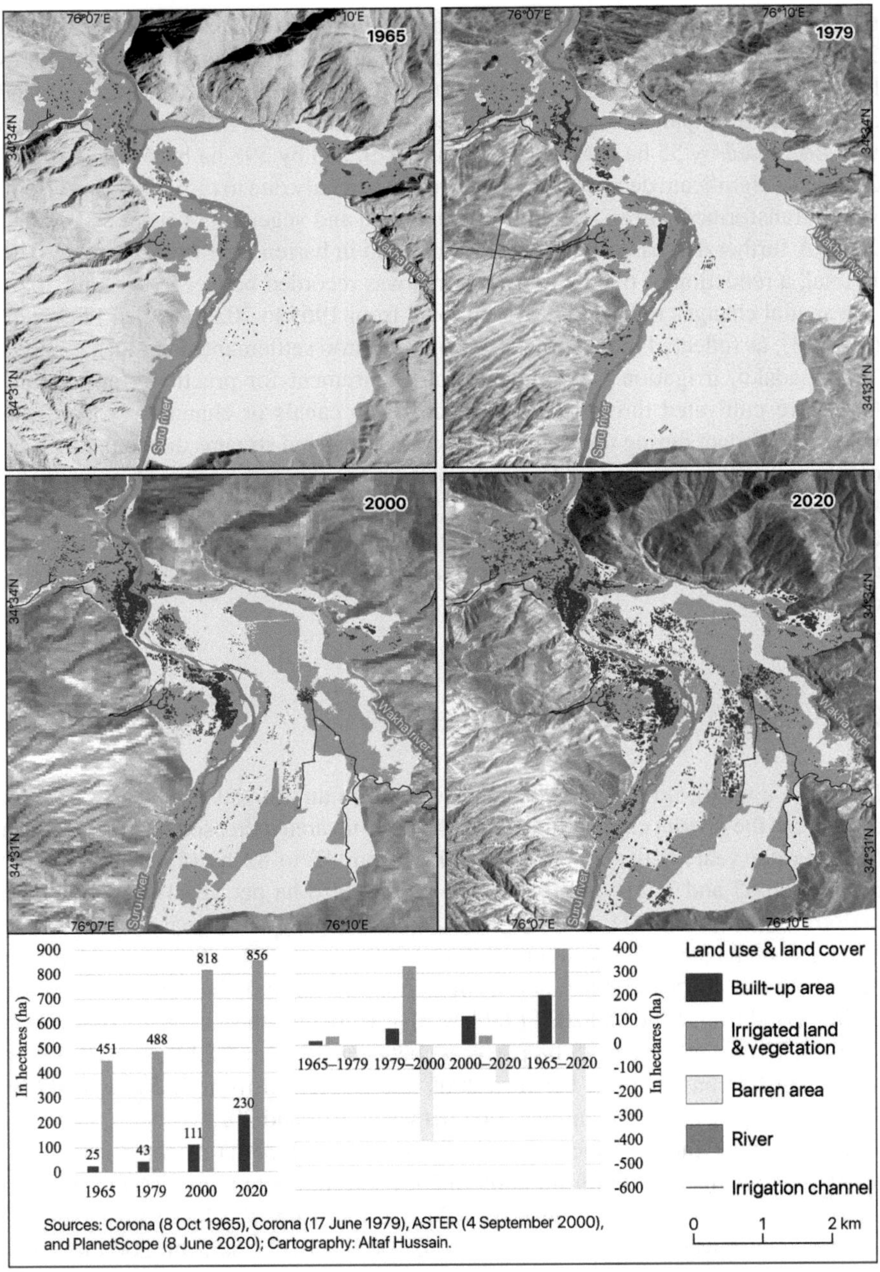

Fig. 7.3 Spatial change of LULC in Kargil town from 1965 to 2020

Fig. 7.4 View of Baroo settlement in Kargil town, Ladakh. The substantial residential area of Baroo is visible as a major structure, combined with commercial establishments along the bank of the Suru River. Photograph taken by Altaf Hussain, August 2023

Table 7.2 LULC and its change detection from 1980 to 2021 in Padum

Area change (ha)	Years	Class		
		Built-up area	Irrigated land and vegetation	Barren area
Area	1980	10	742	5157
	2005	45	869	4995
	2021	82	990	4837
Area per year	1980–2005	1.4	5.08	−6.48
	2005–2021	2.31	7.56	−9.87
	1980–2021	1.76	6.04	−7.80
Change detection area	1980–2005	35	127	−162
	2005–2021	37	121	−158
	1980–2021	72	248	−320

Sources Satellite imagery: Corona (11 September 1980), ASTER (13 October 2005), and PlanetScope (24 July 2021)

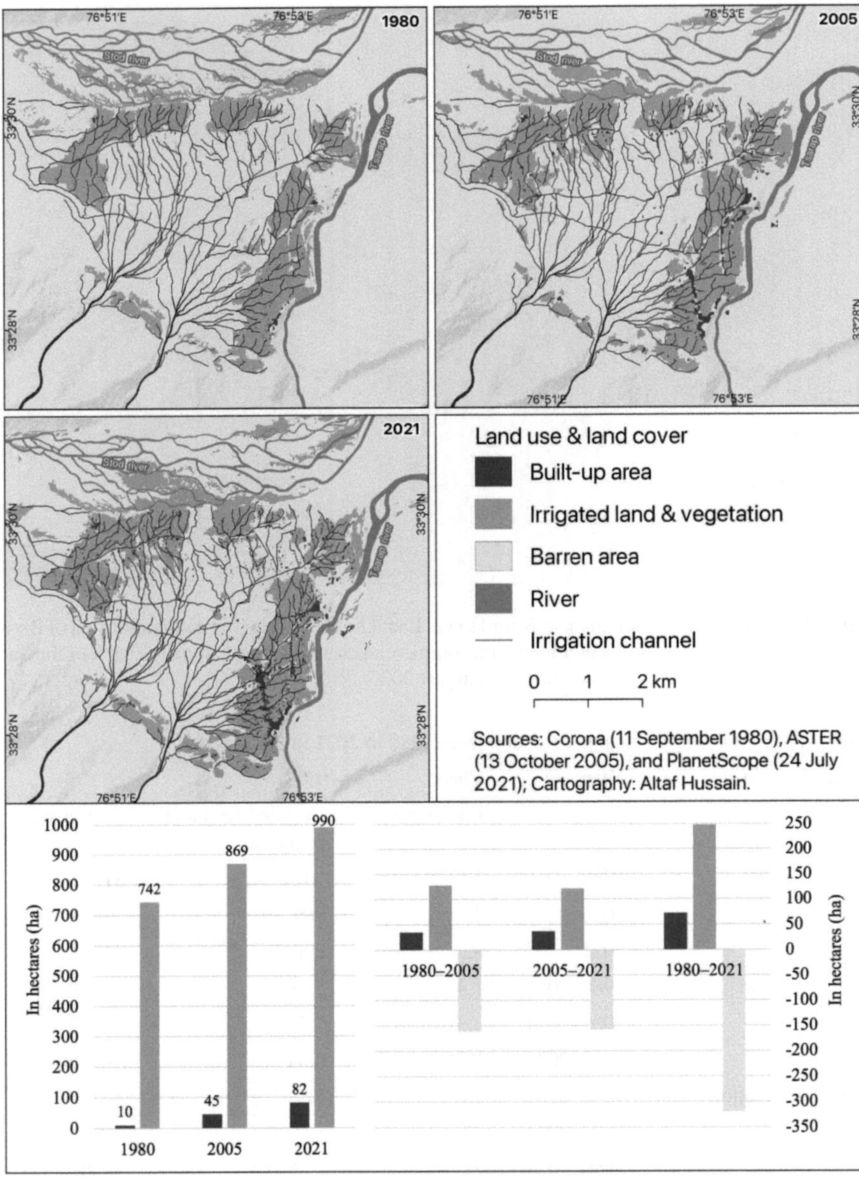

Fig. 7.5 Spatial change of LULC from 1980 to 2021 in Padum

In the 1970s, Padum had a primary and secondary school, a medical centre, a police station, and even a tourist office (Hirschberg, 1978). The total number of inhabitants was 139, with 35 households in 1981 (Khan, 1981). By 2011, the number of inhabitants had increased to 1,538, residing in 261 households (District Statistical & Evaluation Office Kargil, 2020). Today, Padum offers a wide range of services,

Fig. 7.6 Built-up area and irrigated land in Padum. The central area of Padum is visible, functioning as the main hub for surrounding hamlets and villages in the Zanskar region. Photograph taken by Altaf Hussain, August 2022

including a hospital, a degree college, a tourist reception centre, schools, hotels, guesthouses, administrative offices, restaurants, and local retail shops. The expanding supply capacity in the surrounding area is evident, supported by the existing infrastructure. Thus, Padum, as the tehsil administrative centre, plays a crucial role in meeting the daily needs and providing essential commodities to the rural population (Fig. 7.6).

Padum was formerly the capital as well as the trade (tsonglen) centre of Zanskar and the obvious road terminus from Kargil (Crook & Osmaston, 1994; Gazetteer of Kashmir and Ladák, 1890). The impact of industrial technology and commercialisation was seen throughout the vast region of Zanskar, which began to change the traditional socio-economic customs and practices (Crook & Osmaston, 1994). Apart from educational and medical facilities, the government-based food stores were among the most important developmental measures, which brought new food and fuel commodities (Crook & Osmaston, 1994). Following the construction of the road connecting Zanskar and Kargil, there was a noticeable rise in the cash economy. Agricultural practices also experienced substantial changes, influenced by external factors such as government subsidies on staple foods, rural development programmes, and the emergence of non-agricultural income sources (Dame & Mankelow, 2010; Mankelow, 2003). The emergence of tourism led to the widespread monetisation of the economy and multiplied the demand for industrial commodities (Crook & Osmaston, 1994). Furthermore, the connectivity of the NH-301 and the NPD road has brought significant transformation in recent years, leading to more interactions with the outside world in term of trade, commerce, and tourism.

Since 1976, agricultural control plots demonstrating new techniques, fertilisers, and vegetables varieties have been introduced, along with the establishment of veterinary centres. It is estimated that households in central Zanskar grow around 60% of their own food, while the remaining 40% is obtained through purchases from ration stores, private vendors in the Padum bazaar, and, to a lesser extent, from other local households in Zanskar (Dame & Mankelow, 2010). There is also a clear tendency towards livelihood diversification, increased market integration, and external development interventions in the adjacent mountain valleys of Zanskar (Dame & Mankelow, 2010). For instance, in the rural village (yul) of Stongde in Zanskar, livelihoods have been modified due to external interventions, such as the public distribution system[2] (agricultural subsidises), which have impacted traditional household food systems and agricultural practices (Dame & Mankelow, 2010).

7.1.3 LULC Change in Sankoo

Sankoo is a small urbanised settlement located in the Suru Valley of Kargil district. The built-up area increased from 10 to 39 ha (1 ha per year) between 1971 and 2000, from 39 to 65 ha (1.24 ha per year) between 2000 and 2021, and from 10 to 65 ha (1.10 ha per year) between 1971 and 2021. The built-up area expanded more than sixfold over the observation period from 1971 to 2021. The irrigated land and vegetation showed a gradual increase from 746 to 817 ha (2.45 ha per year) between 1971 and 2000, from 817 to 861 ha (2.09 ha per year) between 2000 and 2021, and from 746 to 861 ha (2.30 ha per year) between 1971 and 2021. The barren area decreased from 1,736 to 1,636 ha (3.45 ha per year) between 1971 and 2000, from 1,636 to 1,566 ha (3.33 ha per year) between 2000 and 2021, and from 1,736 to 1,566 ha (3.40 ha per year) between 1971 and 2021. The LULC of Sankoo has been analysed for the period from 1971 to 2021 (Table 7.3).

The built-up area in Sankoo increased by 29 ha between 1971 and 2000 and by an additional 26 ha between 2000 and 2021, resulting in a total increase of 55 ha in built-up area over the period from 1971 to 2021. Similarly, irrigated land and vegetation increased by 71 ha from 1971 to 2000 and by 44 ha from 2000 to 2021, leading to a total increase of 115 ha in irrigated land and vegetation over the entire period from 1971 to 2021. The barren area saw a decrease of 100 ha from 1971 to 2000 and 70 ha from 2000 to 2021, resulting in a total decrease of 170 ha in barren area from 1971 to 2021. The spatial change of LULC in Sankoo from 1971 to 2021 can be observed in Fig. 7.7.

[2] The public distribution system refers to a network of retail outlets (ration stores) through which the Indian government sells subsidised rice, wheat, and kerosene at fixed prices lower than those of the market value. During the 1960s, the public distribution system was introduced in Ladakh as

Table 7.3 LULC and its change detection from 1971 to 2021 in Sankoo

Area change (ha)	Years	Class		
		Built-up area	Irrigated land and vegetation	Barren area
Area	1971	10	746	1736
	2000	39	817	1636
	2021	65	861	1566
Area per year	1971–2000	1	2.45	−3.45
	2000–2021	1.24	2.09	−3.33
	1971–2021	1.10	2.30	−3.40
Change detection area	1971–2000	29	71	−100
	2000–2021	26	44	−70
	1971–2021	55	115	−170

Sources Satellite imagery: Corona (28 September 1971), ASTER (18 July 2000), and PlanetScope (12 August 2021)

In the 1970s, Sankoo was the largest settlement, with 180 houses, a mosque, a school, a fish farm, and many shops in the entire Suru Valley (Hirschberg, 1978; Mattausch, 1993). Presently, the urbanised settlement has a degree college, a hospital, a tourist bungalow, schools, restaurants, guesthouses, government offices, and commercial shops. The total number of households in 1981 was 254, with 1,389 inhabitants (Khan, 1981). By 2011, Sankoo had increased to 349 households and 2,698 inhabitants, covering an area of 448 ha (District Statistical & Evaluation Office Kargil, 2020). In addition, there are 20 inhabited villages in Sankoo tehsil (District Statistical & Evaluation Office Kargil, 2021) (Fig. 7.8).

Since 1978, the Suru Valley has been accessible by a motorable road (Grist, 2008). Since then, the valley has experienced significant land-use change, settlement growth, and economic development. The NH-301 is the primary road, which passes through the middle of Sankoo. Most of the residential areas have developed along the roads and are surrounded by irrigated land, including vegetation and a few orchards. Every household practices farming. The majority of the irrigated land in Sankoo is used for cropping, particularly wheat (kro) and barley (nas). However, a drastic growth of settlements has been observed around the Sankoo bazaar in recent times. The Suru Valley is predominantly agricultural, with staple crops such as barley (nas), wheat (kro), peas, and a few vegetables. Extensive pastures support various livestock, including bovines, horses, sheep, and goats (Grist, 1998, 2008). The construction of road, the opening of tourism, and government and army activities have brought about enormous economic changes (Grist, 2008).

a welfare intervention, and the number of ration stores and food imports increased considerably in the 1990s, with 100% coverage by the late 2000s (Dame, 2023).

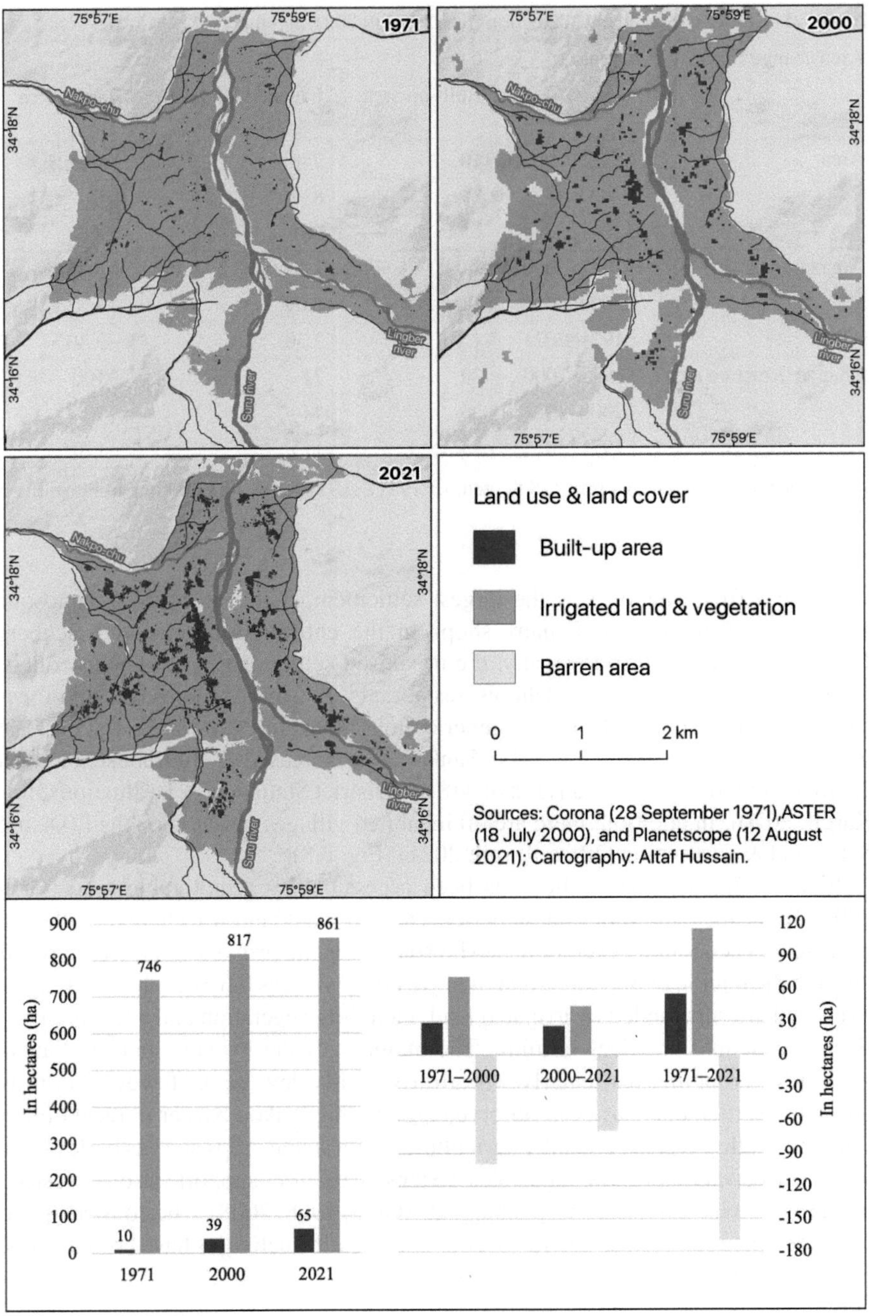

Fig. 7.7 Spatial change of LULC from 1971 to 2021 in Sankoo

Fig. 7.8 Built-up area and cultivated land in Sankoo. The urbanised settlement is surrounded by plantations dominated by poplars, willows, and fruit trees in the Suru Valley. Photograph taken by Altaf Hussain, September 2022

7.1.4 LULC Change in Drass

Drass is a small urbanised settlement and hill station located in the western portion of Kargil district. In Drass, the built-up area increased from 12 to 37 ha (0.71 ha per year) between 1965 and 2000, from 37 to 85 ha (2.28 ha per year) between 2000 and 2021, and from 12 to 85 ha (1.30 ha per year) between 1965 and 2021. At the same time, the irrigated land and vegetation increased from 457 to 607 ha (4.28 ha per year) between 1965 and 2000, from 607 to 654 ha (2.29 ha per year) between 2000 and 2021, and from 457 to 654 ha (3.52 ha per year) between 1965 and 2021. The barren area decreased from 1,703 to 1,528 ha (5 ha per year) between 1965 and 2000, from 1,528 to 1,433 ha (2.28 ha per year) between 2000 and 2021, and from 1,703 to 1,433 ha (1.30 ha per year) between 1965 and 2021. The LULC change of Drass from 1965 to 2021 is shown in Table 7.4.

The built-up area increased by 25 ha between 1965 and 2000, followed by an increase of 48 ha between 2000 and 2021. Consequently, a total built-up area of 73 ha was observed from 1965 to 2021 in Drass. Additionally, irrigated land and vegetation increased by 150 ha between 1965 and 2000, and by 47 ha between 2000 and 2021, resulting in a total increase of 197 ha between 1965 and 2021. The barren area decreased by 175 ha between 1965 and 2000 and by 95 ha between 2000 to 2021, resulting in a total decrease of 270 ha of barren area between 1965 and 2021. The

Table 7.4 LULC and its change detection from 1965 to 2021 in Drass

Area change (ha)	Years	Class		
		Built-up area	Irrigated land and vegetation	Barren area
Area	1965	12	457	1703
	2000	37	607	1528
	2021	85	654	1433
Area per year	1965–2000	0.71	4.28	5
	2000–2021	2.28	2.29	4.52
	1965–2021	1.30	3.52	4.82
Change detection area	1965–2000	25	150	−175
	2000–2021	48	47	−95
	1965–2021	73	197	−270

Sources Satellite imagery: Corona (8 October 1965), ASTER (18 July 2000), and PlanetScope (18 July 2021)

spatial changes in Drass from 1965 to 2021 are evident, demonstrating significant transformations in LULC over time (Fig. 7.9).

In 1981, there were a total of 132 households with 599 inhabitants living in Drass (Khan, 1981). By 2011, the urbanised settlement grew to 404 households and 8,397 inhabitants (District Statistical & Evaluation Office Kargil, 2020). Drass now includes a degree college, a hospital, schools, a tourist bungalow, hotels and guest-houses, administrative offices, army buildings, restaurants, and many retail shops. NH-1 passes from the middle part of this hill station, which has contributed to the expansion of Drass bazaar along the NH-1. The area of the bazaar has grown over the past twenty years, particularly following the Kargil War in 1999 (Fig. 7.10).

In Ladakh, including Kargil, agriculture combined with animal husbandry forms the cornerstone of livelihood and food security. In addition to practicing of agriculture, animal husbandry plays a crucial role in supplying meat, milk products, and providing food in rural areas. Livestock also provide manure (lut) to households during the cropping season for cultivated fields (jhing). Agriculture and animal husbandry have been significant in generating moderate income, particularly for rural households. Food items such as rice (bras), vegetables, pulses, tea (cha), and sugar (khara) are mostly purchased from local bazaars and government ration shops. In urban areas, agriculture and animal husbandry are increasingly shifting towards non-agrarian activities. Agricultural fields and vegetation rely on human-made reservoirs (rzing) and irrigation channels (yourba) for irrigation due to insufficient rainfall (charpha).

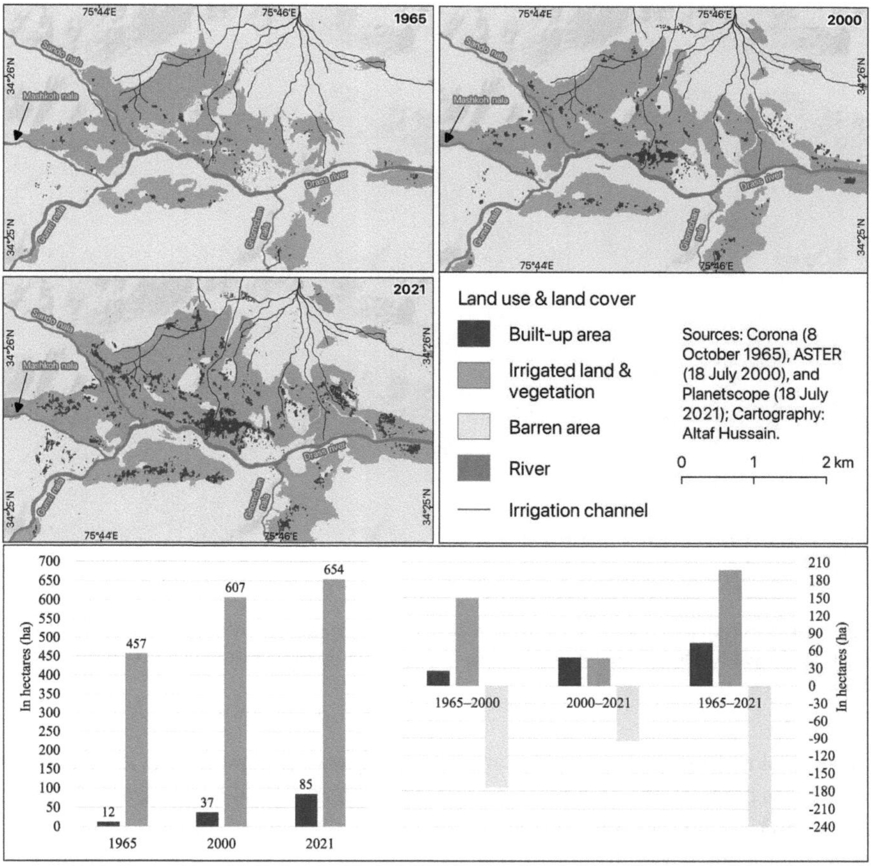

Fig. 7.9 Spatial change of LULC from 1965 to 2021 in Drass

7.2 Occupational Distribution of Workers

The occupational distribution of main and marginal workers from 1981 to 2011 in Kargil town and the urbanised settlements of Padum, Sankoo, and Drass are depicted in Fig. 7.11. The number of main workers in Padum increased from 90 in 1981 to 249 in 2011. While no marginal workers were recorded in Padum in 1981, this number rose to 39 by 2011. Among main workers in Padum, the number of cultivators saw a sharp decline from 85 in 1981 to just 4 in 2011, whereas the number of other workers increased substantially from 4 to 231 during the same period.

In Sankoo, the number of main workers declined from 688 in 1981 to 372 in 2011, while the number of marginal workers rose from 152 to 319 over the same period. Within the category of main workers in Sankoo, there was a notable decrease in the number of cultivators from 469 in 1981 to 35 in 2011, whereas the number of other workers rose from 179 to 292 during the same period. Additionally, Drass saw a

Fig. 7.10 Built-up area and irrigated land in Drass. The tree plantations are visible from the cultivated land used for crop farming in the Drass Valley. Photograph taken by Altaf Hussain, August 2022

substantial rise in the number of main workers, growing from 161 in 1981 to 6,966 in 2011. In contrast, the number of marginal workers increased more gradually, from 223 to 563 during the same period. Among the main workers in Drass, the number of cultivators increased from 24 to 60 between 1981 and 2011, while the number of other workers increased from 137 to 6,902 during the same period.

Moreover, the number of main workers in Kargil town increased significantly from 1,440 to 6,102 between 1981 and 2011, while the number of marginal workers increased from 404 to 1,311 over the same period. Among the main workers in Kargil town, the number of cultivators decreased from 324 to 62 between 1981 and 2011, while the number of other workers increased from 1,082 to 6,010 during the same period.

The occupational distribution of the number and percentage of main and marginal workers from 2001 to 2011 in the blocks of Kargil, Zanskar, Sankoo, and Drass is shown in Fig. 7.12. It can be observed that the number of main workers in Kargil block decreased from 7,930 (73%) to 3,378 (50.3%) between 2001 and 2011, whereas the number of marginal workers increased from 2,936 (27%) to 3,337 (49.7%) during the same period. In Kargil block, the number of cultivators among main workers decreased from 2,195 (20.2%) in 2001 to 417 (6.2%) in 2011, while the number of other workers showed a declining trend from 5,564 (51.2%) to 2,886 (43%) between 2001 and 2011.

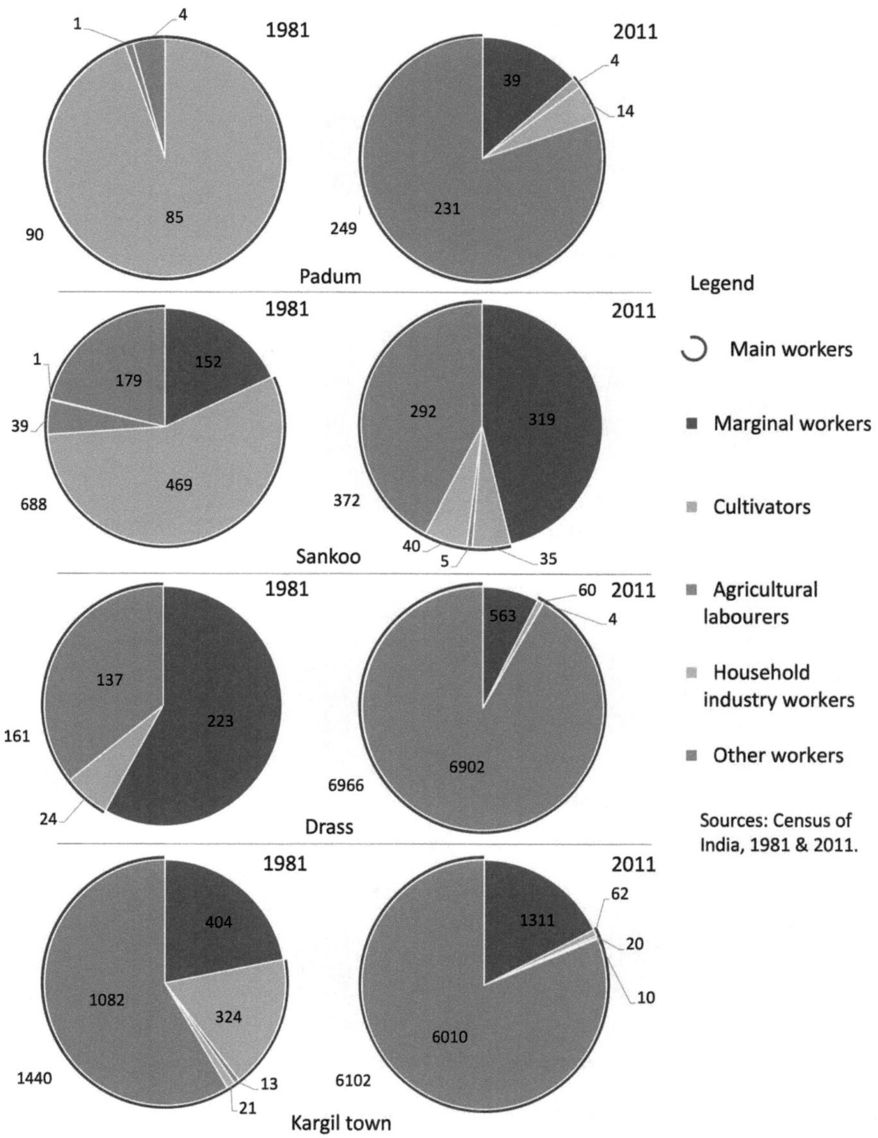

Fig. 7.11 Occupational distribution of workers from 1981 to 2011 in Kargil town and the urbanised settlements

The total number of main workers in Zanskar block decreased from 4,622 (95.3%) to 2,113 (40.3%) between 2001 and 2011, while the number of marginal workers increased from 230 (4.7%) to 3,127 (59.7%) during the same period. In the category of main workers in Zanskar block, a decrease in cultivators was observed, from 2,476 (51%) in 2001 to 482 (9.2%) in 2011, while other workers were shown a declining trend, from 2,130 (44%) to 1,306 (24.9%) between 2001 and 2011.

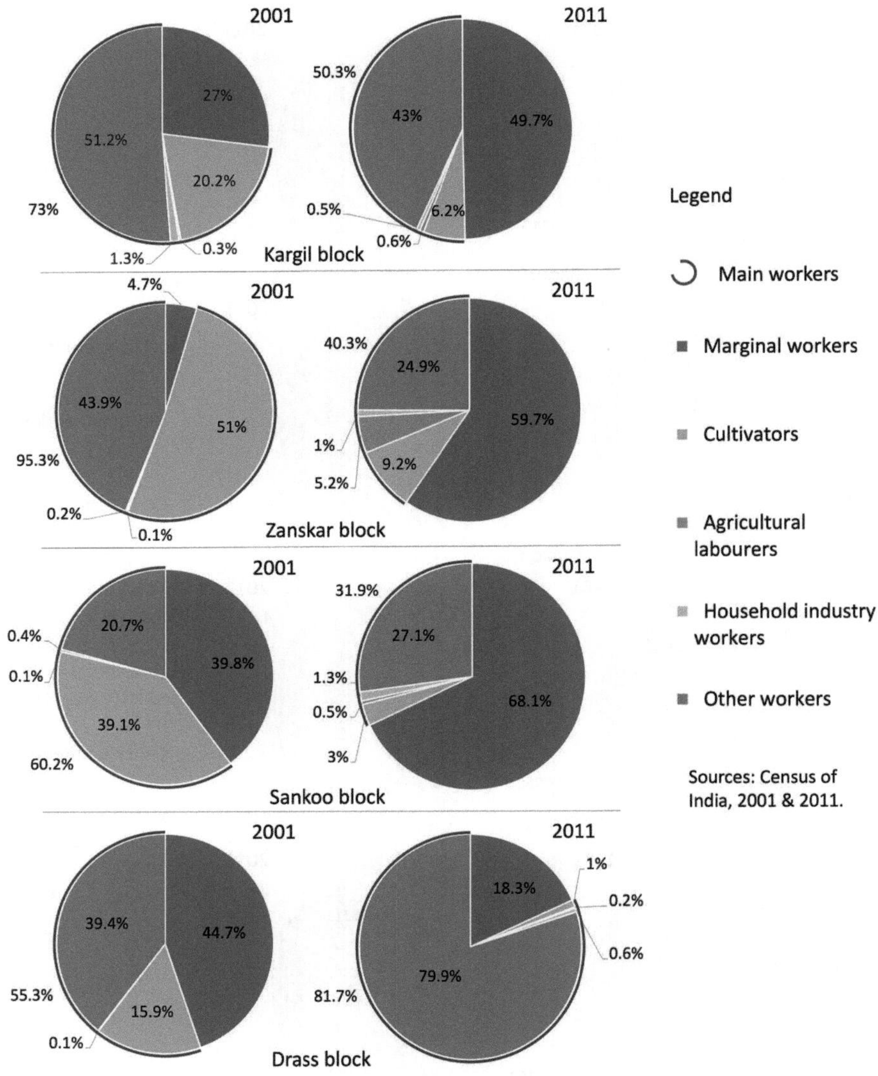

Fig. 7.12 Occupational distribution of main and marginal workers in the blocks from 2001 to 2011

Similarly, it can also be observed in Sankoo block that the number of main workers decreased from 4,505 (60.2%) to 1,666 (31.9%) between 2001 and 2011, while the number of marginal workers increased from 2,974 (39.8%) to 3,550 (68.1%) during the same period. The main workers of Sankoo block witnessed a sharp decline in cultivators, from 2,925 (39.1%) in 2001 to 158 (3%) in 2011.

In addition, it can be observed that the number of main workers increased significantly from 3,136 (55.3%) to 8,608 (81.7%) in Drass block between 2001 and 2011, while the number of marginal workers decreased slightly from 2,533 (44.7%) to

1,933 (18.3%) during the same period. In the category of main workers in Drass block, a decrease in the number of cultivators decreased from 899 (15.9%) to 109 (1%) between 2001 and 2011, while the number of other workers increased from 2,232 (39.4%) to 8,423 (79.9%) during the same period.

In the Kargil district, the distribution of the main workers is mostly concentrated in agricultural activities and other workers, with a sharp decline in the number of cultivators. Many cultivators are moving away from farming and transitioning into other workers (engaged in economic activities). Consequently, there has been a shift from agricultural to non-agricultural activities due to rapid urbanisation. Significant transitions have been observed from subsistence agriculture towards income-based opportunities, public distribution system, and livelihood diversification (Dame, 2018; Dame & Nüsser, 2011).

Ladakh has undergone overwhelming transformations in the political, economic, and social spheres over the past decades. These changes have significantly impacted local livelihoods and raised question about sustainable development (Dame & Nüsser, 2008). Key factors such as border disputes, military presence, government programmes and subsidies, and off-farm income sources have played a prominent role in shaping these changes (Dame & Nüsser, 2008). The transition to a primarily cash-based economy, significantly shaped by the presence of the military, rapid administrative development, and the growth of tourism, has deeply influenced the remote mountainous areas. These transformations have not only increased the material welfare of people in larger towns and villages but also elevated the standard of living in more isolated regions (Fox et al., 1994). In addition to these changes, the increasing rural-to-urban migration, driven by urbanisation, has modified family relationships and socio-economic structures in mountain villages, particularly among nomadic communities (Goodall, 2004). Furthermore, the animal husbandry practices of agropastoral communities have undergone notable transformations due to socio-economic and geopolitical changes in Ladakh (Ladon et al., 2023). These combined factors reflect a broader change in local livelihoods in response to rapid urbanisation and external pressures.

References

Crook, J., & Osmaston, H. (Eds.). (1994). *Himalayan Buddhist villages: Environment, resources, society and religious life in Zangskar, Ladakh* (1st ed.). Motilal Banarsidass Publishers Private Limited.

Dame, J. (2018). Food security and translocal livelihoods in high mountains: Evidence from Ladakh, India. *Mountain Research and Development, 38*(4), 310. https://doi.org/10.1659/MRD-JOURNAL-D-18-00026.1

Dame, J. (2023). Changing production, changing consumption: Food system transformation in Ladakh. In B. Humbert-Droz, J. Dame, & T. Morup (Eds.), *Environmental change and development in Ladakh, Indian Trans-Himalaya* (pp. 175–195). Springer Nature Switzerland. https://doi.org/10.1007/978-3-031-42494-6_12

Dame, J., & Mankelow, J. S. (2010). Stongde revisited: Land-use change in Central Zangskar. *Erdkunde, 64*(4), 355–370. https://doi.org/10.3112/erdkunde.2010.04.05

Dame, J., & Nüsser, M. (2008). Development perspectives in Ladakh, India. *Geographische Rundschau International Edition, 4*(4), 20–27. https://www.researchgate.net/profile/Marcus-Nuesser/publication/279175071_Dame_und_Nuesser_2008_GRIntern/links/558bb5ea08ae02c9d1f968c3/Dame-und-Nuesser-2008-GRIntern.pdf

Dame, J., & Nüsser, M. (2011). Food security in high mountain regions: Agricultural production and the impact of food subsidies in Ladakh, Northern India. *Food Security, 3*(2), 179–194. https://doi.org/10.1007/s12571-011-0127-2

District Statistical & Evaluation Office Kargil. (2020). *Village amenity directory*. Directorate of Economics & Statistics, Planning Development & Monitoring Department.

District Statistical & Evaluation Office Kargil. (2021). *Statistical handbook*. Directorate of Economics & Statistics, Planning Development & Monitoring Department.

Fox, J. L., Nurbu, C., Bhatt, S., & Chandola, A. (1994). Wildlife conservation and land-use changes in the Transhimalayan region of Ladakh, India. *Mountain Research and Development, 14*(1), 39. https://doi.org/10.2307/3673737

Gazetteer of Kashmir and Ladák. (1890). *Together with routes in the territories of the Maharája of Jamú and Kashmír; compiled under the direction of the quarter master general in India in the intelligence branch*. Superintendent of Government Printing, India.

Goodall, S. K. (2004). Rural-to-urban migration and urbanization in Leh, Ladakh: A case study of three nomadic pastoral communities. *Mountain Research and Development, 24*(3), 220–227. https://doi.org/10.1659/0276-4741(2004)024[0220:RMAUIL]2.0.CO;2

Grist, N. (1998). *Local politics in the Suru Valley of Northern India*. Goldsmiths, University of London. https://research.gold.ac.uk/id/eprint/11450/1/272599.pdf

Grist, N. (2008). Urbanisation in Kargil and its effects in the Suru Valley. In M. van Beek, & F. Pirie (Eds.), *Modern Ladakh: Anthropological perspectives on continuity and change*. Brill. https://doi.org/10.1163/ej.9789004167131.i-313

Hill, J. (2014). Farmer-managed irrigation systems in Baltistan and Kargil. *Ladakh Studies, 31*, 4–23.

Hirschberg, H. (1978). *Ladakh, das andere Tibet: Mit Zanskar*. Geobuch-Verlag.

Khan, A. H. (1981). *District census handbook, Kargil district: Village and town directory, series-8, part XIII-A & B*. Director of Census Operations, Jammu & Kashmir. https://censusindia.gov.in/nada/index.php/catalog/29355

Khan, K. A. (2020). The unmaking of Kargil town. *Stawa*. https://stawa.org/the-unmaking-of-kargil-town/

Ladon, P., Nüsser, M., & Garkoti, S. C. (2023). Mountain agropastoralism: Traditional practices, institutions and pressures in the Indian Trans-Himalaya of Ladakh. *Pastoralism, 13*(1), 30. https://doi.org/10.1186/s13570-023-00289-1

LAHDC-Kargil. (2016). *District irrigation plan (Kargil district) under Pradhan Mantri Krishi Sinchayee Yojana (PMKSY)*. Government of Jammu and Kashmir.

Mankelow, J. S. (2003). *The implementation of the watershed development programme in Zangskar, Ladakh: Irrigation development, politics and society*. University of London. https://www.soas.ac.uk/water/publications/papers/file38409.pdf

Mattausch, J. (1993). *Ladakh und Zanskar: Handbuch für individuelles Entdecken im äußersten Norden Indiens* (1st ed.). Reise Know-How Verlag Peter Rump Gmbh.

Nüsser, M., Schmidt, S., & Dame, J. (2012). Irrigation and development in the Upper Indus Basin: Characteristics and recent changes of a socio-hydrological system in Central Ladakh, India. *Mountain Research and Development, 32*(1), 51–61. https://doi.org/10.1659/MRD-JOURNAL-D-11-00091.1

Rizvi, J. (2012). *Ladakh: Crossroads of High Asia* (3rd ed.). Oxford University Press.

Zutshi, J. N., Farooqi, B. A., Din, G. M., & Nanda, J. K. (1961). *Census of India: Village survey monograph of Kharboo (Tehsil Kargil, District Ladakh), Part VI, No. 13, Vol. VI*. Director of Census Operations, Jammu & Kashmir.

Chapter 8
Discussion, Conclusion, and Recommendations

Abstract This chapter presents the empirical research findings and compares them with other scientific studies conducted in Ladakh as well as in the broader Trans-Himalayan region. The comparison aims to highlight similarities, differences, and unique insights, enhancing the overall understanding of the urbanisation processes. It synthesises empirical research to enhance understanding of urban development challenges, specifically in Kargil, Ladakh. By integrating these empirical findings, it emphasises the evolving patterns of urban growth and development, offering insights into how socio-economic factors, environmental effects, and cultural aspects collectively influence its development path. Moreover, the chapter proposes recommendations to address urban and environmental challenges, advocating for practical measures aligned with urban planning norms, regulations, and policy frameworks. Urgent measures are needed, including improved governance, infrastructure development, and sustainable tourism policies to ensure inclusive and resilient urban growth in this mountainous region. Finally, the chapter concludes by acknowledging its limitations and suggests future research directions for exploring new areas of investigation.

Keywords Urbanisation · Urban development · Rural-to-urban migration · Road connectivity · Tourism · Infrastructure development · Urban planning · Sustainable development · Kargil · Ladakh

8.1 Discussion

In the Trans-Himalaya, Ladakh has been a subject of interest for researchers, social scientists, and policymakers across various disciplines. The region's fragile ecosystem, characterised by cold, high-altitude deserts, has faced numerous social and environmental challenges, including climate change, habitat degradation, permafrost melting, geopolitical conflicts, and external interventions. The improved accessibility of the region through motorised transport has facilitated the development of market centres, tourism activities, and infrastructure. The introduction

A. Hussain, *Urbanisation and Development Challenges in Kargil, Ladakh*, Advances in Asian Human-Environmental Research, https://doi.org/10.1007/978-3-032-05667-2_8

of road connectivity in 1960 and the development of tourism in 1974 have been major drivers of rapid socio-economic and environmental transformations. Significant progress has been made in constructing and upgrading road infrastructure, enhancing connectivity across the region. This development has been largely driven by geopolitical factors, including the region's proximity to international borders and the need for secure transportation routes. The evolving political landscape, along with the growing importance of roads for economic growth, tourism, and military logistics, has further accelerated infrastructure modernisation. Furthermore, the formation of LAHDC-Leh in 1995 and LAHDC-Kargil in 2003 has contributed to infrastructure development and the transformation of the landscape structure. This political shift in governance has not only facilitated increased financial investment but also further accelerated infrastructure development, amplifying the region's socio-economic and environmental transformation. After Ladakh was granted UT status in 2019, the district annual capex budget increased from approximately ₹57 crores to more than ₹300 crores in both LAHDC-Kargil and LAHDC-Leh. This financial empowerment has led to rapid growth and development across sectors such as telecommunication, health care, tourism, education, water infrastructure, electricity and renewable energy, and transportation. As a result, the region has undergone major transformations due to improved accessibility, tourism, and strengthened local governance structures.

8.1.1 Urban Development in Ladakh

Urban development in Ladakh has accelerated significantly since the region was granted hill council status. The LAHDCs have enabled a more structured approach to urban planning and development, leading to improved infrastructure and urban management. Local governance through LAHDCs has facilitated development in sectors like health care, education, and rural development, along with improvements in road connectivity, energy infrastructure, telecommunications, and other key infrastructure. Tourism plays a key role in economic growth, attracting visitors with the region's natural beauty and cultural heritage. Urban growth in Leh town has been largely influenced by infrastructure development, the tourism industry, its role as an administrative hub, the emergence of urban culture, and the region's strategic geopolitical position (Dame et al., 2019). Consequently, Leh has transformed into a key point of exchange, attracting a diverse mix of visitors and traders, including international tourists and merchants from Kashmir and other parts of India, as well as local traders (Müller & Dame, 2016). The town's expansion began in 1973 with the establishment of educational institutions, administrative offices, military facilities, and infrastructure projects on previously barren land. The earliest military installations, including the airport, were constructed in the town's south-western area, initiating the first wave of development (Müller, 2022). From the 1990s onwards, new neighbourhoods such as Housing Colony, Skalzangling, and Ibex Colony rapidly grew along the southern periphery of Leh. By 2003, the development of additional residential and administrative areas led to a sharp increase in the built-up environment, with the number of buildings rising to approximately 11,800 (Dame et al., 2019).

In Chitral town, urban development was evident through the expansion of its bazaar, residential zones, and modern built-up areas equipped with both public and private services. This growth was primarily driven by regional population increases, the extension of government administration, and economic influences such as the rise of mountain tourism (Dittmann & Nüsser, 2002; Nüsser, 2001). Similarly, in neighbouring Tibet, Lhasa experienced substantial urban expansion fuelled by government policies, economic progress, tourism growth, and rising investments. Between 1990 and 2008, a series of government-led initiatives promoted infrastructure development in Lhasa, with urban growth intensifying between 2008 and 2015 due to rapid economic expansion, a sharp increase in tourist inflows, and continued government support. Tourism emerged as a major driver of economic development and urban sprawl, with the introduction of the Tibetan railway further accelerating urbanisation (Tang et al., 2017). In contrast, urban growth in the Kathmandu Valley was shaped by a combination of physical geography, accessibility, economic prospects, land market dynamics, population increase, political context, and various development plans and policies (Thapa & Murayama, 2010). This growth resulted in a rapid decline of agricultural land and a significant expansion of built-up areas, largely fuelled by rural-to-urban migration (Ishtiaque et al., 2017; Muzzini & Aparicio, 2013). However, the lack of effective urban planning and weak regulatory enforcement led to uncontrolled urban sprawl, dramatically altering the urban landscape of the valley (Muzzini & Aparicio, 2013). Likewise, several regions within the HKH have experienced rapid, unregulated urbanisation due to rising population pressures and the absence of effective land-use policies (Rasul et al., 2019).

Kargil has witnessed profound changes over the past few decades, particularly following significant events such as the 1999 conflict. This period marked a turning point, as the conflict acted as a catalyst for numerous developmental initiatives aimed at enhancing various aspects of life in the region. The 1999 conflict brought about a massive development drive, which led to notable improvements in economic growth, basic education, infrastructure, telecommunication, and tourism (Milton, 2013). It can also be inferred from the results that the region has experienced rapid demographic and socio-economic changes. In Kargil town, the main drivers shaping and transforming the urban landscape include population growth, rural-to-urban migration, the tourism sector, urban expansion, and historical development (Hussain et al., 2023a). These factors have placed significant pressure on existing urban infrastructure, leading to an increase in critical urban issues such as solid waste management, water supply demands, inadequate deep drains, insufficient toilets, traffic congestion, and a lack of public spaces. Unplanned urban development and encroachment on agricultural land, barren areas, riverbanks, and slopes have further intensified the critical urban challenges in Kargil town (Hussain et al., 2023a).

The urban development of Leh is shaped by the Leh Municipal Committee, LAHDC-Leh, Indian army, and NGOs (Dame et al., 2019). As part of the UIDSSMT scheme, substantial infrastructure developments have been carried out, such as the provision of household water supply, the installation of the town's first sewage system, and the construction of a sewage treatment facility (Dame et al., 2019). Additionally, under the Rajiv Awas Yojana (RAY), efforts have been made to incorporate vernacular architecture into new buildings, install streetlights, and enhance sanitation to improve living conditions and foster tourism development (Müller & Dame, 2016). However, the rapid growth of tourism in Leh has placed considerable strain on water resources and sanitation systems. Soak pits, commonly used by private households, public buildings, and tourist facilities, are often located near borewells, raising significant concerns about water quality. The growing reliance on soak pits becomes more pronounced with the increased water usage driven by the rise in guesthouses and hotels, as well as the installation of more bathrooms in private homes. Guesthouses and hotels produce far more wastewater than private households, contributing to the escalating demand for wastewater disposal solutions (Müller et al., 2020). The growth of the tourism sector, along with the increasing use of flush toilets and showers, has significantly raised both water consumption and wastewater generation (Gondhalekar & Akhtar, 2015). This surge in water usage, coupled with a departure from traditional dry toilets, has contributed to a rise in untreated wastewater, primarily due to the absence of adequate sewage treatment infrastructure (Müller, 2022). In Leh town, the shift towards modern sanitation is evident in the widespread adoption of flush toilets and the construction of soak pits for wastewater disposal, replacing the conventional dry sanitation systems (Dolma et al., 2020). The influx of tourists and migrants has played a key role in driving this transition, leading to increased groundwater extraction through borewells (Choral, 2018). Traditional dry toilets have long served as an important method of soil enrichment and remain the primary means of human waste disposal in most villages across Ladakh, with the exception of Leh and Kargil towns (Bhasin, 2002). For instance, in the old town of Leh, all the toilets have a traditional composting system, which is emptied manually when full (used as fertiliser in agricultural fields), and is disposed of at a dumping site due to the decline of agricultural fields (Alexander, 2007). In the modern town of Leh, toilets with septic tanks have become very popular, as all the hotels and guesthouses have water-flush septic tanks, which have aggravated the water shortages for the urban population (Alexander, 2007).

Moreover, in Ladakh, SBM has been implemented by both the MCs of Leh and Kargil since 2014 to achieve universal sanitation coverage and safe sanitation. In the 4,041 statutory towns, approximately 7.9 million households lack access to toilets and continue to engage in open defecation (Ministry of Housing & Urban Affairs, 2017). Since the establishment of the UT of Ladakh in 2019, the H&UDD is the nodal department for ensuring the balanced and planned growth of urban areas with adequate policies and urban planning. While the department is not yet fully functional, it has implemented various projects in Ladakh's urban areas, including the provision of safe drinking water, sanitation, stormwater drainage, sewerage, roads, street lighting, public transport, and affordable housing.

8.1.2 Rapid Urbanisation in Ladakh

Urbanisation in Ladakh is marked by a decline in agricultural practices in rural areas, primarily due to the attraction of urban centres, resulting in a shortage of male labour in the villages (Nüsser et al., 2012). The government developmental initiatives and the influence of tourism have triggered significant transformations in the socio-economic landscape, transitioning the traditional economy towards livelihood diversification and monetisation (Crook et al., 1994). The livelihood diversification involves transitioning away from agriculture to pursue job opportunities available in the nearby larger towns. While agriculture remains important, most households now rely primarily on non-agricultural income sources. Leh town, functioning as the central hub for services such as health care, education, commerce, and administration, plays a pivotal role in attracting rural populations (Dame, 2018; Dame et al., 2019). This dynamic has led to seasonal increases in urban populations, resulting in labour shortages and socio-cultural disruptions in rural areas. At the same time, agricultural production has declined, increasing reliance on market supplies and food subsidies (Dame, 2018). As a result, the burden of work has increased for women and elderly family members who remain in rural homes. Over recent decades, Leh town has undergone considerable socio-economic transformation, which has notably altered both its physical layout and the social and material character of the town. One significant transformation has been the gradual breakdown of Leh's traditional division into distinct agricultural, commercial, and residential areas (Müller, 2022). Urbanisation is now largely driven by spatial densification and the expansion of the built-up area, influenced by population growth, social change, and especially the growth of the tourism industry (Dame et al., 2019). Migration from rural areas to Leh has become a central feature of this process, spurring the development of new neighbourhoods on the town's periphery. People typically move for reasons such as family reunification, better employment, access to education, and the desire for an urban lifestyle (Müller, 2022).

Urbanisation has emerged as one of the significant drivers of urban development and transformation. In recent years, rapid urbanisation has led to the intervention of remote and inaccessible areas due to improved road connectivity, the growth of the tourism sector, and globalisation. Kargil town, for example, has experienced substantial migration from nearby villages, putting considerable pressure on land and infrastructure (Town Planning Organisation, 2018). Similarly, Leh town experiences major seasonal shifts, with its population rising significantly during the summer months due to tourists, migrant labourers, and defence personnel (Dame et al., 2019). The level of urbanisation in Leh has increased from 12.7 to 23.4% between 1981 and 2001, showing a higher level of urbanisation compared to Kargil (Goodall, 2004). This is reflected in the town's total population, which surged from 3,546 to 30,870 between 1951 and 2011 (Dame et al., 2019; Müller, 2022). Moreover, India has observed rapid growth of urbanisation over the last few decades. The share of the urban population has risen from 23% in 1980 to 32% in 2011 and to 34% in 2018, with the total population growing over threefold since 1950 to reach 1.35 billion

(United Nations, 2019; World Bank, 2013). Approximately half of the country's 285 million urban residents live in towns with fewer than 100,000 people, and many of these smaller towns are experiencing population growth rates comparable to those of larger cities (UN-HABITAT, 2010). Between 2018 and 2050, India is expected to add 416 million more urban residents, almost doubling its urban population (United Nations, 2019).

The Himalayas are currently the most rapidly urbanising mountain range in the world, experiencing swift yet largely unregulated urban growth and expansion. Over recent decades, the region has witnessed significant urban growth, primarily driven by population increase, improved road infrastructure, expanding tourism, and the forces of economic globalisation (Tiwari & Joshi, 2020). In Kargil, rural-to-urban migration is influenced by employment opportunities, access to education, business prospects, healthcare services, and various other opportunities in Ladakh. These factors are supported by additional research focusing on the dynamics of urbanisation across rural hinterlands and mountain towns in the HKH (Dame et al., 2019; Mukherji et al., 2018; Singh et al., 2020; Tiwari et al., 2018). Many areas within the HKH have experienced fast demographic and economic growth, which has raised the demand for natural resources and often led to their overexploitation (Wang et al., 2019). This trend has led to significant LULC change, habitat fragmentation, and unsustainable socio-economic activities. In the HKH, many sustainability challenges stem from escalating demands on natural resources, exacerbating environmental degradation, uncontrolled urbanisation, infrastructural development, and the erosion of traditional culture (Wang et al., 2019; Wester et al., 2019).

Over the past few decades, Ladakh has experienced persistent growth and development in its tourism sector, making it one of the fastest-growing industries in the region. In Kargil, the post-1999 Kargil War period and the establishment of LAHDC-Kargil in 2003 marked a turning point, leading to significant advancements in road infrastructure, communication systems, and various development initiatives. These improvements have contributed to a notable increase in the influx of both domestic and international tourists, driven by the expansion of tourism-related activities. The construction of tourist assets like hotels, guesthouses, and travel agencies has multiplied, particularly in the last two decades. Most of the hotels, guesthouses, restaurants, and travel agencies are located in the main bazaar of Kargil, Baimathang, Baroo, and the Kurbathang Plateau. Affluent residents are primarily engaged in the construction of hotels, guesthouses, and restaurants. The Zoji La tunnel is set to become India's longest road tunnel and will provide all-weather connectivity between Kashmir and Ladakh. The completion of the Zoji La tunnel will bring further interventions and reduce travel time for passengers, tourists, and travellers travelling by road. For centuries, the local communities maintained a largely self-sufficient lifestyle, relying on subsistence agriculture, livestock rearing, and the traditional caravan trade for their livelihoods. The region has gradually become more exposed to external influences, fostering processes of modernisation and globalisation through government-led programmes and development initiatives. This shift has led to an increased dependence on both the lowland Indian economy and global economic systems (Beek & Pirie, 2008; Dame & Nüsser, 2008, 2011). Increasing tourism, population growth, military presence, and the extensive influx of imported raw materials, goods, and commodities represent key dimensions of recent human interventions in the region (Pelliciardi, 2013, 2021). In recent decades, tourism in Ladakh has expanded considerably, leading to substantial economic and social transformation. However, this rapid growth has also resulted in negative environmental consequences. The fragile ecosystem of the region is increasingly at risk due to insufficient infrastructure, inadequate planning mechanisms, and the absence of effective policy interventions (Geneletti & Dawa, 2009). Sustainable tourism is essential to minimising environmental impact, support local communities, preserve cultural heritage, promote biodiversity conservation, and ensure the sustainability of tourism activities in Ladakh.

8.1.3 Rural-to-Urban Migration

Rural-to-urban migration has increased rapidly in the capital towns of Ladakh due to administrative, commercial, educational, and employment opportunities. This phenomenon became more evident after Ladakh was granted governance and administrative powers through the formation of LAHDC in Leh in 1995 and in Kargil in 2003. The establishment of the LAHDCs has acted as a major driving force for rural populations to migrate to the administrative towns, seeking better opportunities and an improved quality of life. According to the 2011 Census of India, urban population growth in Leh has been higher compared to Kargil, although most of the population in Ladakh still resides in rural areas. The urban expansion of Leh has been shaped by multiple factors, including natural population growth, rural-to-urban migration, the settlement of Tibetan refugees, the seasonal influx of labourers, and the arrival of traders from other parts of Jammu and Kashmir (Goodall, 2004, 2007). The opening of borders for tourism has accelerated human impact on the fragile environment since the 1970s, particularly from outside borders. This has led to a growing influx of Indian migrants and has drawn rural inhabitants of Ladakh towards Leh, the region's administrative centre (Dame et al., 2019; Goodall, 2004; Müller & Dame, 2016).

For example, in response to rapid urban growth and the influx of seasonal migrants, new residential wards have been developed in areas such as Ibex and Housing Colony since the 1990s (Müller & Dame, 2016). Migrants are primarily employed in both skilled and unskilled sectors, including roles within the Indian army, government services, and small-scale businesses (Goodall, 2004). However, the rural population has declined as younger generations have increasingly migrated to Leh in pursuit of employment and educational opportunities (Norberg-Hodge, 2009). In recent years, intensive construction activity has occurred both in the central areas and on the outskirts of Leh, reflecting a diversification of urban functions and services (Dame et al., 2019). The growth of commercial enterprises, tourism infrastructure, and road networks has not only reshaped the central bazaar but has also significantly altered the town's spatial structure over the past few decades (Müller &

Dame, 2016). The increasing rural-to-urban migration has modified socio-economic characteristics and family relationships in mountain villages particularly in nomadic communities (Goodall, 2004). Subsistence agriculture, the main economic mainstay, is shifting towards income-based opportunities, leading to a reduction in agricultural workers on farmlands. Farmers, who once mutually helped each other without money exchange, now must pay agricultural workers and labourers (Norberg-Hodge, 1992, 2009). Government-led road construction initiatives have contributed to an increasing reliance on subsidised goods and services among communities in Ladakh (Dame & Mankelow, 2010). Likewise, in the high-altitude Hunza Valley, external interventions have transformed subsistence-based livelihoods, leading to a shift towards non-agricultural sources of income (Kreutzmann, 1993).

In addition, Kargil town has witnessed immense rural-to-urban migration from various rural areas of Kargil district. Baroo Colony, Poyen Colony, Andoo Colony, and the settlements in Kurbathang Plateau are mostly comprised of migrants from Chiktan, Sodh, Drass, Suru, and other rural areas. Haidery Mohalla consists of people who migrated from Drass, while the inhabitants of Silmo Colony came from Silmo village, which is close to the LOC in the Batalik sector. Barchey and Tumail Colonies in Kurbathang are predominantly inhabited by migrants from Sodh, highlighting the ongoing rural-to-urban migration in the area. The growth of these settlements is driven by improved economic prospects, enhanced infrastructure, and greater access to essential services in Kargil town. Besides the local migrants, there are seasonal migrants coming into Kargil from outside Ladakh. The seasonal migrants are mostly shopkeepers from Kashmir, carpenters from Punjab, labourers from Doda in Jammu, workers from Bihar and Uttar Pradesh, and Nepali labourers from Nepal. Mostly, the seasonal migrants live in Kabadi Lungma in Changchik, Poyen Colony, the old bazaar area, and Baroo Colony. These migrants have increased the demand for town services and infrastructure, including water supply, sewage systems, waste disposal, and hygiene facilities. Moreover, the seasonal migration of local people from Kargil to Kashmir, Jammu, Delhi, and other states has increased rapidly. Well-off families have purchased land and constructed houses, leading to the establishment of colonies in various pockets of Kashmir and Jammu. The major driving factors have been the aspiration for a better standard of living, improved health care, business opportunities, employment, and education.

8.1.4 Road Connectivity

When Srinagar-Kargil-Leh road was completed and opened to light vehicular traffic in August 1960 (Zutshi et al., 1961), Ladakh, including Kargil, was no longer an inaccessible area to the rest of India. This road played a crucial role in enhancing accessibility to the region and facilitating economic, social, and strategic interactions with the rest of the country. The Sino-Indian War in 1962 and territorial conflicts with Pakistan led to the construction of the NH-1 and the NH-3 between the 1960s and 1970s (Baghel & Nüsser, 2015; Demenge, 2011; Kreutzmann, 1991; Zutshi et al., 1961). After the Kargil War of 1999, the central government approved 28 road projects under various schemes for different parts of the Kargil in 2005 (Milton, 2013). Due to the geopolitical situation, road infrastructure in the Ladakh region has increased, particularly by the Border Roads Organisation (BRO) and the National Highways and Infrastructure Development Corporation Limited (NHIDCL).

The NH-1 has strategic importance and is closed for vehicular traffic due to heavy snowfall in the Zoji La pass during winter for around 4 months. The completion of Asia's longest Zoji La tunnel (14 km) will provide all-weather road connectivity between Kashmir and Ladakh. The Suru Valley has been accessible via a motorable road since 1978 (Grist, 2008). Moreover, the road (NH-301) from Kargil passes through Padum via Suru Valley and Penzi La, representing the sole motorable road since the 1970s (Milton, 2013). Since then, there has been tremendous land-use change, settlement growth, and economic development. The NH-301 is being widened and upgraded by NHIDCL. The completion of the Zoji La tunnel and NH-301 will be crucial steps towards improving road connectivity and boosting economic growth in the region. As an alternative road to Zanskar, the BRO started construction of the Chadar road from Nimmu to Padum in 2000 to provide road connectivity between Zanskar and Leh (Milton, 2013). Recently, the opening of the Padum to Darcha road for vehicular traffic has reduced the distance and travel time between the Zanskar Valley and the Lahaul and Spiti district. The infrastructure improvements to the NPD road are expected to bring more external intervention, particularly after the completion of the Shinku La tunnel (4.1 km). NPD is a 297 km stretch of road connecting the UT of Ladakh and the state of Himachal Pradesh.

Moreover, NH-3 connects Leh, the capital of Ladakh, with Manali in Himachal Pradesh. The Kargil to Skardu route, once a key historical connection between Ladakh and Baltistan, has remained closed since the 1947 partition of India and Pakistan. The improvement of road infrastructure has been a significant priority of the central government due to the threat posed by China at the borders of Ladakh, as well as ongoing border conflicts with Pakistan. To enhance connectivity and boost economic development, the Indian government has made substantial investments in road infrastructure, focusing on widening and strengthening existing roads, constructing new bridges, and improving road safety measures. In December 2023, the central government sanctioned ₹1,170.16 crore for 29 road development projects in the UT of Ladakh, focusing on the construction and improvement of state highways, major roads, and district roads. Additionally, ₹181.71 crore was approved for eight bridge projects under the Central Road and Infrastructure Fund (CRIF) scheme for the 2023–24 financial year. The district capital expenditure budget of LAHDCs of Kargil and Leh has recently increased sixfold, reflecting significant overall development in Ladakh. For the financial year 2024–25, LAHDC-Kargil has proposed a budget of approximately ₹363 crore, marking the highest district capital expenditure budget since the formation of the autonomous body.

Ladakh's air connectivity is crucial because of the region's strategic and geopolitical significance, a fact emphasised during the 1962 Sino-Indian War. The Kushok Bakula Rimpochee Airport, established in Leh in 1985, remains the key air facility for Ladakh, continuing to meet the region's essential transportation needs. It has played a crucial role in supporting both civilian and military operations, significantly improving access to this remote region. To further enhance infrastructure and passenger services, the expansion of Leh's airport has been planned with an estimated cost of ₹480 crores, ultimately aiming to transform it into a world-class facility (Times of India, 2020). Additionally, the terminal will accommodate more than 200 cars and is expected to increase the airport's annual passenger capacity from 900,000 to 2 million. This development is part of broader efforts to improve transportation infrastructure in Ladakh. The expansion is anticipated to significantly boost tourism, trade, and economic development in the region (Times of India, 2020).

Kargil's connectivity relies heavily on road networks, especially NH-1, which is crucial for accessing Srinagar to the west and Leh to the east. This highway acts as the town's main connection to the wider region, enabling the movement of essential goods, military equipment, and tourist traffic. While Kargil has a small airport, it is primarily used for military operations, with no civilian flights available at present. Consequently, both travellers and businesses are entirely dependent on road transportation, which intensifies the difficulties related to connectivity. To address these challenges, initiatives are being taken to enhance the airport's capacity to support civilian flights. This upgrade would mark a significant improvement in regional connectivity, positioning Kargil as a more prominent hub and enabling greater mobility for residents, tourists, and businesses. Enhanced air transport options would not only improve Kargil's accessibility but also strengthen the Ladakh region's integration with the rest of India. The enhanced connectivity would drive economic growth, create educational opportunities, and promote tourism in the region, ultimately contributing to the overall progress of Ladakh.

8.1.5 Opening of Tourism Sector

Tourism has developed drastically in the mountain regions of the Trans-Himalaya, resulting in massive economic, social, and environmental changes. When the tourism sector was opened to foreign tourists in 1974, it brought tremendous changes in the socio-economic conditions of the people of Ladakh. Despite political instability in Kashmir, the number of tourist arrivals multiplied from 25,000 in 1988, 28,400 in 2003, 79,100 in 2010, and 327,366 in 2018 (Dame et al., 2019; Geneletti & Dawa, 2009; Pelliciardi, 2021). The rapid growth of tourism has made it the predominant service sector in Ladakh, generating revenue and creating employment opportunities (Pelliciardi, 2010). As tourism expanded, labour migration also rose, especially from Nepal and the Indian lowlands. Most of these jobs are seasonal, given that tourist visits are largely limited to the summer months (Dame et al., 2019). The dependence on the modern wage economy, and the increased interactions with Western culture through tourism have transformed mountain communities (Michaud, 1996; Norberg-Hodge, 2009). In recent years, agricultural land throughout Ladakh has increasingly been converted for commercial use, including the development of guesthouses, restaurants, and retail establishments (Clouse, 2021). In Leh, the number of guesthouses and hotels increased from 24 in 1980 to 360 in 2013. However, the rapid expansion of tourist accommodations has led to the widespread installation of bathrooms with flush toilets and showers, even though a proper sewage system is still lacking (Gondhalekar & Akhtar, 2015).

Additionally, Pangong Tso, the most popular tourist destination in the Changthang region, has witnessed a substantial increase in tourism-related activities. Until 2001, overnight stays near the lake were prohibited. However, by 2008, the number of visitors had risen sharply to an estimated 100,000 to 150,000, resulting in the establishment of 21 permanent camps, six hotel-resorts, 30 homestay guesthouses, and 17 café-restaurants. Similarly, the Tsokar-Tsomoriri region remained inaccessible to tourists until the 1990s. By 2018, the area had developed approximately 25 homestay guesthouses, 10 semi-permanent camps, eight campsites, seven hotels and resorts, and several café-restaurants, catering to an estimated 80,000 visitors (Humbert-Droz, 2023). However, this rapid tourism expansion has also contributed to environmental degradation, especially in the Hemis and Tsokar-Tsomoriri National Parks. The fragile ecosystem of Ladakh is under increasing pressure due to inadequate infrastructure, lack of proper policies, and insufficient planning tools (Geneletti & Dawa, 2009). Tourism development, while boosting the economy, poses a significant threat to environmental conservation. Good road connectivity and commercial flights have made Leh more accessible, facilitating the movement of both tourists and residents. In contrast, Kargil remains reliant on the Srinagar-Kargil-Leh road, making road transport the primary means of access for travellers. Kargil's airport is currently used only for military purposes and remains closed to commercial flights. Additionally, the closed border with Pakistan and the winter closure of road connectivity have hindered tourism growth and limited development opportunities in Kargil compared to Leh in Ladakh.

8.1.6 Increase of Natural Hazards

Kargil has a limited area for urban expansion and development. Due to this constraint, most buildings are constructed on barren land and steep slopes, making them vulnerable to natural hazards. In recent years, flash floods have repeatedly impacted the region, with major events in 2022 affecting Baroo and Titichumik in Kargil town, as well as rural villages in Chiktan, Suru, and Drass. These floods caused extensive damage to roads and agricultural areas, highlighting the growing risks posed by extreme weather events (Hussain et al., 2023b). The region remains at risk of future flash floods triggered by cloudbursts, which may become more frequent due to regional climate change. Studies suggest that increasing variability in precipitation patterns could further exacerbate these hazards (Chevuturi et al., 2018; Nüsser et al., 2022; Schmidt & Nüsser, 2012, 2017; Schmidt et al., 2020). The flash flood that struck Leh in 2010 stands as a recent instance of a natural disaster that resulted in considerable loss of human lives and public property (Le Masson, 2015; Thayyen et al., 2013).

Over the past two decades, both Leh and Kargil have experienced multiple flash flood events (Leh in 2006, 2008, 2010, 2015, and 2018; Kargil in 2006, 2010, 2013, 2014, 2016, 2018, and 2022), underscoring the increasing vulnerability of Ladakh to these natural disasters (Dame et al., 2019; Hussain et al., 2023b). A major concern in urban planning is the growing exposure to natural hazards, which is intensified by construction in vulnerable zones such as alluvial fans and riverbanks—areas highly prone to flash floods and debris flows during intense rainfall (charpha) events (Dame et al., 2019). The relationship between mountain development and disaster risk is intrinsic, as many mountain communities are situated on unstable slopes susceptible to landslides and erosion, or on river terraces and alluvial fans that are particularly vulnerable to floods and debris flows (Vaidya et al., 2019). Climate-related hazards, including flash floods and landslides, have significantly contributed to the rise in disasters affecting both mountainous regions and downstream areas, putting increasing pressure on local populations (Adler et al., 2022). Kargil and its adjoining valleys face significant threats from natural hazards like landslides, flash floods, avalanches, and earthquakes, endangering the lives of residents in both the town and surrounding areas. To reduce these risks, construction should be restricted in zones vulnerable to such hazards, particularly areas prone to flash floods and landslides. Identifying vulnerable disaster-prone zones and taking appropriate measures through the District Disaster Management Authority (DDMA) of LAHDC-Kargil is essential. Furthermore, implementing building bylaws that restrict construction in disaster-prone areas and ensuring their strict enforcement would contribute to improving safety and resilience in the region.

8.2 Conclusion

Ladakh is undergoing rapid urbanisation, with urban areas emerging as significant drivers of economic growth. However, in the capital towns of Leh and Kargil, this urbanisation is unfolding in a largely haphazard manner, accompanied by extensive infrastructure development across the region. This process is marked by a decline in agricultural activities in rural areas, primarily driven by the growth of Leh's urban centre, which has brought about profound socio-economic changes over recent decades (Dame et al., 2019; Goodall, 2004; Müller, 2022; Nüsser et al., 2012). The transformation in settlement patterns began after the opening of the NH-1 for motorised vehicles in 1960. Improved accessibility, the construction of roads, and the growing number of motor vehicles have significantly altered movement patterns in the region, while the road's access to irrigated land has facilitated the construction of new houses, leading to urban sprawl. This expansion has resulted in the development of homes on steep slopes and barren land. Additionally, the demand for housing, driven by settlement growth and demographic pressure, has resulted in the conversion of irrigated land into residential areas, further impacting agricultural productivity and land-use patterns.

In the case of Kargil town, the rural-to-urban migration has become a predominant force driving urban change, with a noticeable increase in migration over time. The number of migrants moving from rural to urban areas for economic reasons is steadily increasing, with many choosing to settle in the administrative centre of Kargil town. This migration trend is contributing to the town's rapid urbanisation, placing additional pressure on its infrastructure, housing, and public services as the population grows. The main reasons for rural-to-urban migration include employment, education, business, healthcare, and other opportunities. In recent years, agricultural land has decreased due to encroachment from commercial, administrative, and residential activities. Locals often construct residential and commercial buildings on their agricultural land, particularly along roads, generating revenue by renting them out. Most of the land in Kargil is owned by the local population. However, there is little oversight or control over the conversion of agricultural land to non-agricultural uses, particularly by the MC and the KDA. To avoid uncontrolled expansion, proper demarcation of town boundaries and zoning of existing land use are essential. Land-use zoning can help preserve property values and enable the town to grow inclusively and sustainably. Drafting bylaws for land-use management is needed. Proper, systematic town planning/urban planning is also necessary for Kargil town's resilience, liveability, and sustainable development.

Addressing unplanned development requires integrated approaches that balance economic development with environmental conservation, social equity, and cultural preservation. This involves strengthening governance mechanisms, enhancing community participation in decision-making processes, promoting sustainable land-use planning and management practices, and investing in resilient infrastructure and services that can address the unique challenges of mountain environments. In Kargil, urban infrastructure is under significant strain due to the rising urban population and rural-to-urban migration. As the primary hub of economic activity, the town has experienced unplanned development, resulting in rising solid waste generation, inadequate water supply, traffic congestion, sanitation issues, substandard drainage systems, and a notable absence of appropriate recreational amenities and public spaces. These urban development challenges demand urgent attention and integrated solutions. By integrating environmental sustainability, social equity, and community participation into urban planning and development, the town can move toward a more resilient, inclusive, and sustainable model of mountain development that benefits both present and future generations.

Furthermore, Kargil faces distinctive challenges concerning LULC transformation within the broader context of Ladakh. Rapid population growth and urban expansion have led to the transformation of natural landscapes into built-up areas, including residential, commercial, and public infrastructure. The observed LULC changes include the conversion of irrigated land and vegetation into built-up areas, the transformation of barren land into both agricultural and built-up areas, and a significant overall increase in built-up areas. The Ladakh region has recently experienced significant transformations due to driving factors such as tourism, military activities, infrastructure development, urbanisation, and environmental changes. Tourism and road connectivity have emerged as significant drivers of change and have shaped

LULC patterns. The attraction of mountain landscapes, adventure activities, and cultural experiences has rapidly increased tourism infrastructure, tourist arrivals, and associated development. This influx of tourists brings opportunities and challenges for mountain communities and their cultural heritage in Ladakh. Tourism-driven development can stimulate economic growth, create employment opportunities, and support local businesses in Ladakh. However, unmanaged tourism activities can exert pressure on the fragile mountain environment, leading to habitat degradation and loss of biodiversity. Traffic congestion, waste generation, toilets, and water scarcity are common issues associated with excessive tourism development in Ladakh. Moreover, the commodification of local culture and landscapes can result in cultural homogenisation, loss of traditional practices, and social disruptions within local communities. Sustainable tourism practices and responsible management strategies are essential to addressing these challenges. This includes promoting ecotourism initiatives that prioritise environmental conservation, cultural preservation, and community engagement.

8.3 Recommendations

Addressing the challenges of Kargil requires integrated and holistic approaches that prioritise inclusive and sustainable development in Ladakh. In Kargil town, adequate wastewater treatment systems are necessary to treat effluents, as all wastewater is currently disposed of untreated directly into the Suru River. Commercial activities such as vehicle workshops, service stations, and butcher shops should be discouraged from locating near the Suru River to prevent further environmental degradation. The LAHDC-Kargil must regulate wastewater from residential, commercial, and public buildings while ensuring the provision of adequate treatment facilities. The disposal of solid waste into the Suru River should be strictly banned, with fines imposed for violations. The solid waste treatment facility at Kurbathang Plateau should be properly managed, and policies should be formulated for the allocation of waste disposal sites, including those for hospital waste. The urban drains constructed by LAHDC-Kargil are primarily designed for stormwater collection and are not intended for municipal greywater disposal. Existing roadside drains lack the capacity to handle larger volumes of stormwater during the rainy season, leading to frequent blockages in both open and closed drainage systems. The town's domestic water supply is sourced from water channels, public tube wells, piped networks, and water tankers. However, rising urban populations, rural-to-urban migration, and increasing construction activities have led to higher water demand, affecting both quality and availability. Municipal services need to be strengthened across all urban wards to improve accessibility and enhance the overall liveability of Kargil town.

Kargil lacks a proper public transport system. The existing parking stands in the town are inadequate for serving the local population and people commuting from rural areas. The increasing number of private and official vehicles during rush hours leads to severe traffic congestion. Due to inadequate parking facilities, many vehicles are parked along the roadsides, further narrowing the lanes available for movement. There are only a few footpaths in the town centre, and none are pedestrian friendly. The narrow roads in the town centre should be converted into pedestrian and bicycle-friendly zones. Additionally, vehicle-free zones should be designated, where the entry of motorised vehicles is restricted based on traffic conditions. The absence of proper parking areas, roadside encroachments, lack of traffic signals, inadequate road upgrades, and weak enforcement of regulations have worsened traffic congestion in Kargil town. Measures should be taken to encourage walking and cycling in the old town centre and commercial areas while discouraging haphazard roadside parking. Strict regulations should be implemented to prevent unauthorised modifications to streets and roads, with penalties for violations. LAHDC-Kargil should formulate bylaws to regulate heavy traffic within the town by setting specific time slots for vehicular movement. Constructing city bypasses can help reduce traffic on internal roads. Additionally, vehicles travelling to other parts of Kargil district should be assigned designated routes to minimise congestion within the town.

The Suru River, which flows through the central part of Kargil town, has significant potential for recreational development and revenue generation. Converting the riverbank into recreational zones and developing public spaces can significantly enhance the quality of life for urban residents while also boosting tourism growth. Recreational facilities such as walking and jogging tracks, biking trails, parks, camping sites, and school outing areas can be developed to make the riverbank a vibrant public space. However, unplanned urban development along the Suru River poses serious risks, including human casualties, property damage, and economic disruption. The increasing pressure of unregulated construction on the riverbank requires urgent intervention. Residential, commercial, and public infrastructure developments have led to extensive encroachment, further destabilising the riverbank. Additionally, the riverbanks in Andoo and Baroo Tsok are being extensively exploited for sand and gravel mining. These activities are accelerating the degradation of the riverbanks, bringing the Suru River closer to existing structures and increasing the risk of erosion and flooding. Immediate measures are needed to regulate development, prevent encroachment, and implement sustainable riverbank protection strategies.

Kargil is experiencing unplanned growth and uncontrolled construction across all urban settlements. The lack of proper urban planning, absence of norms and regulations, and indiscriminate use of land for residential, commercial, public, and tourism purposes have contributed to haphazard development. LAHDC-Kargil, along with the MC, is responsible for various development works, including road construction, water supply, municipal solid waste management, electricity provision, sanitation facilities, wastewater treatment, public transport, public spaces, and the enforcement of municipal by laws. However, the town requires well-planned physical and social infrastructure to support sustainable socio-economic development. To address these challenges, LAHDC-Kargil must acknowledge and respond to the towns' urban development issues by creating a comprehensive blueprint. A systematic approach is necessary to manage the growing pattern of unregulated and haphazard development, with integrated urban planning being essential for safeguarding the urban environment and ensuring sustainable growth.

As far as tourism is concerned, the sector in Kargil remains relatively underdeveloped compared to Leh. However, the valleys of Zanskar, Drass, Suru, and Chiktan possess significant tourism potential due to their unique physical landscapes, historical importance, and rich cultural heritage. These valleys hold great promise and can be developed for adventure tourism, winter sports, border tourism, heritage tourism, eco-tourism, and religious tourism. For instance, the Mushkoh Valley, Tiger Hill, Tololing, Marpo La, and the Kargil War Memorial can be major tourist attraction places in Drass Valley, while Zanskar and Suru Valleys can become major tourism hubs for skiing, rafting, mountaineering, trekking, motorcycle tours, forts, traditional houses, and religious tourism. The LAHDC-Kargil and UT administration need to create sustainable development plans at the local, district, and regional levels. Sustainable tourism guidelines should be developed in collaboration with local people and stakeholders, and policies for the sustainable development of tourism must be formulated and implemented in Ladakh.

If Ladakh is to be made liveable, inclusive, and sustainable, local communities and stakeholders must be actively involved in decision-making, planning, and the implementation of both local and central programmes. Formulating bylaws, regulations, and development plans is critical to promoting inclusive urban development while preserving cultural heritage and the natural environment. There is an urgent need for the LAHDCs and the UT administration to strengthen frameworks for policy formulation, urban planning, and governance. Both councils, Kargil and Leh must take action regarding the establishment of a town planning organisation in Ladakh, ensuring equitable and inclusive urban development in this region. Fostering inclusive and sustainable development in Ladakh requires a balanced and integrated approach that considers the unique geographical, cultural, and environmental characteristics. Here are key areas and detailed recommendations for comprehensive development:

- Promote investment in renewable energy, particularly solar, wind, and small-scale hydropower, by harnessing the vast solar potential of Ladakh.
- Develop integrated waste management systems that prioritise recycling, composting, and wastewater treatment to protect local water sources from urban runoff.
- Enhance infrastructure for utilities (water, electricity, sanitation) by implementing smart metering and monitoring systems to ensure reliable services.
- Utilise green building technologies that integrate locally sourced materials and traditional construction techniques with modern energy-efficient practices.
- Develop disaster-resilient infrastructure and early warning systems to reduce risks from flash floods, avalanches, and landslides, often triggered by glacial melt, rainfall, and snowfall.
- Promote efficient irrigation and sustainable agriculture to conserve water and strengthen local food production, particularly in the mountainous region where resources are limited and climate-sensitive.
- Protect cultural and historical sites while establishing eco-tourism initiatives and policies to regulate tourist numbers, ensuring conservation for future generations, safeguarding the environment, and preserving local community welfare.

- Ensure access to essential services like health care, education, and emergency services, even in remote areas.
- Improve transportation systems by developing terrain-appropriate public transit networks, enhancing road infrastructure with proper safety measures, and incorporating smart city technologies such efficient public lighting, intelligent traffic management, GPS-based public transport tracking, and smart parking facilities to ensure safe and sustainable mobility.
- Strengthen regional connectivity within Ladakh and with neighbouring areas to boost economic activities and regional integration.
- Create well-designed public spaces with clear guidelines to promote social interaction, physical activity, and community well-being.
- Develop flexible zoning regulations to support mixed-use developments, encourage efficient land use, and foster sustainable, pedestrian-friendly environments.
- Engage local communities in urban planning to address their needs, preserve cultural heritage, and enhance skills in technology, tourism, and sustainability through targeted training programmes.
- To achieve sustainable development, integrated urban planning must be implemented across key sectors such as the environment, economy, energy, transportation, tourism, land use, disaster management, and health care, ensuring efficient resource use and fostering a balanced, resilient future for all.

8.4 Limitations and Way Forward

The significant limitations were geographical remoteness, the COVID-19 pandemic, and political instability, particularly in collecting data from the study area in Ladakh, India. The census in Jammu and Kashmir in 1991 was not conducted due to the region's prevailing political unrest and security concerns during that time. The state witnessed heightened tensions, mainly due to the insurgency and militancy activities, which made it challenging to carry out a comprehensive census operation. Consequently, the Indian government decided not to conduct the census in Jammu and Kashmir in 1991. Similarly, the Census of India for 2021 has not yet been conducted, primarily due to delays caused by the COVID-19 pandemic and political considerations. Despite advances in technology and research methods, many gaps have yet to be identified in achieving a comprehensive understanding and exploration of the region of Ladakh. The need for more historical records and literature concerning Kargil creates gaps in knowledge and impedes thorough research and analysis. Due to its geographical remoteness and vastness, scientific research in the Ladakh region requires substantial financial resources, specialised equipment, and skilled personnel.

Although many studies have recently been carried out on Ladakh's religion, culture, geopolitics, socio-economic conditions, and environmental issues, more scientific research is still needed on these aspects. Urbanisation in Ladakh presents both challenges and opportunities for sustainable development. By adopting an integrated approach to urban and regional planning, the UT administration of Ladakh, LAHDCs, stakeholders, and local communities can effectively navigate the complexities of urbanisation and broader development challenges. Collaboration, innovation, and resilience will be pivotal in charting a path forward towards inclusive and sustainable development in Ladakh. This study supports the urbanisation trends in mountain regions and provides updated assessments and insights into rapid urbanisation processes in the Himalayan region.

References

Adler, C., Wester, P., Bhatt, I., Huggel, C., Insarov, G., Morecroft, M., Muccione, V., & Prakash, A. (2022). Cross-chapter paper 5: Mountains. In H.-O. Pörtner, D. C. Roberts, M. M. B. Tignor, E. Poloczanska, K. Mintenbeck, A. Alegría, M. Craig, S. Langsdorf, S. Löschke, V. Möller, A. Okem, & B. Rama (Eds.), *Climate change 2022: Impacts, adaptation and vulnerability* (pp. 2273–2318). Cambridge University Press. https://doi.org/10.1017/9781009325844.022

Alexander, A. (2007). *Towards a management plan for the old town of Leh: Structuring a plan for the preservation of an endangered townscape and revitalization of traditional social structures.* http://www.tibetheritagefund.org/media/download/leh_conservation_aa.pdf

Baghel, R., & Nüsser, M. (2015). Securing the heights: The vertical dimension of the Siachen conflict between India and Pakistan in the Eastern Karakoram. *Political Geography, 48*, 24–36. https://doi.org/10.1016/j.polgeo.2015.05.001

Bhasin, V. (2002). Traditional knowledge research: Recycling human waste among Ladakhis. *Journal of Human Ecology, 13*(3), 177–180. https://doi.org/10.1080/09709274.2002.11905530

Chevuturi, A., Dimri, A. P., & Thayyen, R. J. (2018). Climate change over Leh (Ladakh), India. *Theoretical and Applied Climatology, 131*(1–2), 531–545. https://doi.org/10.1007/s00704-016-1989-1

Choral, L. (2018, April 15). Unique toilets of Ladakh. *Daily Excelsior*. https://www.dailyexcelsior.com/unique-toilets-ladakh/

Clouse, C. (2021). *Climate-adaptive design in high mountain villages: Ladakh in transition.* Routledge, Taylor & Francis Group.

Crook, J., & Osmaston, H. (Eds.). (1994). *Himalayan Buddhist villages: Environment, resources, society and religious life in Zangskar, Ladakh* (1st ed.). Motilal Banarsidass Publishers Private Limited.

Dame, J., & Nüsser, M. (2008). Development perspectives in Ladakh, India. *Geographische Rundschau International Edition, 4*(4), 20–27. https://www.researchgate.net/profile/Marcus-Nuesser/publication/279175071_Dame_und_Nuesser_2008_GRIntern/links/558bb5ea08ae02c9d1f968c3/Dame-und-Nuesser-2008-GRIntern.pdf

Dame, J. (2018). Food security and translocal livelihoods in high mountains: Evidence from Ladakh, India XE. *Mountain Research and Development, 38*(4), 310. https://doi.org/10.1659/MRD-JOURNAL-D-18-00026.1

Dame, J., & Mankelow, J. S. (2010). Stongde revisited: Land-use change in Central Zangskar. *Erdkunde, 64*(4), 355–370. https://doi.org/10.3112/erdkunde.2010.04.05

Dame, J., & Nüsser, M. (2011). Food security in high mountain regions: Agricultural production and the impact of food subsidies in Ladakh, Northern India. *Food Security, 3*(2), 179–194. https://doi.org/10.1007/s12571-011-0127-2

Dame, J., Schmidt, S., Müller, J., & Nüsser, M. (2019). Urbanisation and socio-ecological challenges in high mountain towns: Insights from Leh (Ladakh), India. *Landscape and Urban Planning, 189*, 189–199. https://doi.org/10.1016/j.landurbplan.2019.04.017

Demenge, J. (2011). *The political ecology of road construction in Ladakh*. University of Sussex. http://sro.sussex.ac.uk/38501/

Dittmann, A., & Nüsser, M. (2002). Siedlungsentwicklung im östlichen Hindukusch: Das Beispiel Chitral Town (North-West Frontier Province, Pakistan). *Erdkunde*, 60–72. https://www.jstor.org/stable/23218604

Dolma, K., Rishi, M. S., & Lata, R. (2020). State of groundwater resource: Relationship between its depth and sewage contamination in Leh town of union territory of Ladakh. *Applied Water Science, 10*(3), 78. https://doi.org/10.1007/s13201-020-1157-8

Geneletti, D., & Dawa, D. (2009). Environmental impact assessment of mountain tourism in developing regions: A study in Ladakh, Indian Himalaya. *Environmental Impact Assessment Review, 29*(4), 229–242. https://doi.org/10.1016/j.eiar.2009.01.003

Gondhalekar, D., & Akhtar, A. (2015). Down the drain or back to the roots? Political ecology of the water-energy-food nexus visualized using GIS in Leh town, Ladakh, India. In K. Goh & E. Chu (Eds.), *Projections 11—Global ecologies: Politics, planning, and design* (pp. 77–96). The MIT Journal of Planning.

Goodall, S. K. (2007). *From plateau pastures to urban fringe: Sedentarisation of nomadic pastoralists in Ladakh, north-west India*. University of Adelaide. https://digital.library.adelaide.edu.au/dspace/bitstream/2440/46332/8/02whole.pdf

Goodall, S. K. (2004). Rural-to-urban migration and urbanization in Leh, Ladakh: A case study of three nomadic pastoral communities. *Mountain Research and Development, 24*(3), 220–227. https://doi.org/10.1659/0276-4741(2004)024[0220:RMAUIL]2.0.CO;2

Grist, N. (2008). Urbanisation in Kargil and its effects in the Suru Valley. In M. van Beek, & F. Pirie (Eds.), *Modern Ladakh: Anthropological perspectives on continuity and change*. Brill. https://doi.org/10.1163/ej.9789004167131.i-313

Humbert-Droz, B. (2023). Environmental change in Ladakh's Changthang: A local, regional and global phenomenon. In B. Humbert-Droz, J. Dame, & T. Morup (Eds.), *Environmental change and development in Ladakh, Indian Trans-Himalaya* (pp. 119–146). Springer Nature Switzerland. https://doi.org/10.1007/978-3-031-42494-6_9

Hussain, A., Schmidt, S., & Nüsser, M. (2023a). Dynamics of mountain urbanisation: Evidence from the Trans-Himalayan town of Kargil, Ladakh, India. *Land, 12*(4), 920. https://doi.org/10.3390/land12040920

Hussain, A., Schmidt, S., & Nüsser, M. (2023b). *Urban landscape change in the Trans-Himalayan town of Kargil, Ladakh, India*. https://doi.org/10.5194/egusphere-egu23-2663

Ishtiaque, A., Shrestha, M., & Chhetri, N. (2017). Rapid urban growth in the Kathmandu Valley, Nepal: Monitoring land use land cover dynamics of a Himalayan city with Landsat imageries. *Environments, 4*(4), 72.

Kreutzmann, H. (1991). The Karakoram highway: The impact of road construction on mountain societies. *Modern Asian Studies, 25*(4), 711–736. https://doi.org/10.1017/S0026749X00010817

Kreutzmann, H. (1993). Challenge and response in the Karakoram: Socioeconomic transformation in Hunza, Northern Areas, Pakistan. *Mountain Research and Development, 13*(1), 19. https://doi.org/10.2307/3673642

Le Masson, V. (2015). Considering vulnerability in disaster risk reduction plans: From policy to practice in Ladakh, India. *Mountain Research and Development, 35*(2), 104–114. https://doi.org/10.1659/MRD-JOURNAL-D-14-00086.1

Michaud, J. (1996). A historical account of modern social change in Ladakh (Indian Kashmir) with special attention paid to tourism. *International Journal of Comparative Sociology, 37*(3–4), 286–300. https://doi.org/10.1163/002071596X00064

Milton, N. (2013). *Kargil: An abode of Mountain Tamers*. Har-Anand Publications Pvt Ltd.

Ministry of Housing & Urban Affairs. (2017). *Guidelines for Swachh Bharat mission-urban*. Government of India. http://swachhbharaturban.gov.in/writereaddata/SBM_GUIDELINE.pdf

Mukherji, A., Scott, C., Molden, D., & Maharjan, A. (2018). Megatrends in Hindu Kush Himalaya: Climate change, urbanisation and migration and their implications for water, energy and food.

In A. K. Biswas, C. Tortajada, & P. Rohner (Eds.), *Assessing global water megatrends* (pp. 125–146). Springer Singapore. https://doi.org/10.1007/978-981-10-6695-5_8

Müller, J., & Dame, J. (2016). Small town, great expectations: Urbanization and beautification in Leh. *South Asia Multidisciplinary Academic Journal, 14*. https://doi.org/10.4000/samaj.4208

Müller, J. (2022). *Urban mountain waterscapes in Leh, Indian Trans-Himalaya: The transformation of hydro-social relations*. Springer International Publishing. https://doi.org/10.1007/978-3-031-18249-5

Müller, J., Dame, J., & Nüsser, M. (2020). Urban mountain waterscapes: The transformation of hydro-social relations in the trans-Himalayan town Leh, Ladakh, India. *Water, 12*(6), 1698. https://doi.org/10.3390/w12061698

Muzzini, E., & Aparicio, G. (2013). *Urban growth and spatial transition in Nepal: An initial assessment*. The World Bank. https://doi.org/10.1596/978-0-8213-9659-9

Norberg-Hodge, H. (1992). Ancient futures—Learning from Ladakh. *India International Centre Quarterly, 19*(4), 3–13. http://www.jstor.org/stable/23004004

Norberg-Hodge, H. (2009). *Ancient futures: Lessons from Ladakh for a globalizing world*. Sierra Club Books.

Nüsser, M., Schmidt, S., & Dame, J. (2022). *Cryosphere changes and local adaptation strategies: Socio-hydrological case studies from the Himalayan region*. https://doi.org/10.5194/iahs2022-661

Nüsser, M. (2001). Understanding cultural landscape transformation: A re-photographic survey in Chitral, eastern Hindukush, Pakistan. *Landscape and Urban Planning, 57*(3–4), 241–255. https://doi.org/10.1016/S0169-2046(01)00207-9

Nüsser, M., Schmidt, S., & Dame, J. (2012). Irrigation and development in the Upper Indus Basin: Characteristics and recent changes of a socio-hydrological system in Central Ladakh, India. *Mountain Research and Development, 32*(1), 51–61. https://doi.org/10.1659/MRD-JOURNAL-D-11-00091.1

Pelliciardi, V. (2010). Tourism traffic volumes in Leh district: An overview. *Ladakh Studies, 26*, 14–23.

Pelliciardi, V. (2013). From self-sufficiency to dependence on imported food-grain in Leh District (Ladakh, Indian Trans-Himalaya). *European Journal of Sustainable Development, 2*(3), 109–122. https://doi.org/10.14207/ejsd.2013.v2n3p109

Pelliciardi, V. (2021). Factors affecting international and national tourist arrivals (1974–2020) in Leh district (UT Ladakh, India). *European Journal of Sustainable Development, 10*(1), 736–736. https://doi.org/10.14207/ejsd.2021.v10n1p736

Rasul, G., Saboor, A., Tiwari, P. C., Hussain, A., Ghosh, N., & Chettri, G. B. (2019). Food and nutrition security in the Hindu Kush Himalaya: Unique challenges and niche opportunities. In P. Wester, A. Mishra, A. Mukherji, & A. B. Shrestha (Eds.), *The Hindu Kush Himalaya assessment* (pp. 301–338). Springer International Publishing. https://doi.org/10.1007/978-3-319-92288-1_9

Schmidt, S., & Nüsser, M. (2012). Changes of high altitude glaciers from 1969 to 2010 in the Trans-Himalayan Kang Yatze Massif, Ladakh, Northwest India. *Arctic, Antarctic, and Alpine Research, 44*(1), 107–121. https://doi.org/10.1657/1938-4246-44.1.107

Schmidt, S., & Nüsser, M. (2017). Changes of high altitude glaciers in the Trans-Himalaya of Ladakh over the past five decades (1969–2016). *Geosciences, 7*(2), 27. https://doi.org/10.3390/geosciences7020027

Schmidt, S., Nüsser, M., Baghel, R., & Dame, J. (2020). Cryosphere hazards in Ladakh: The 2014 Gya glacial lake outburst flood and its implications for risk assessment. *Natural Hazards, 104*(3), 2071–2095. https://doi.org/10.1007/s11069-020-04262-8

Singh, S., Tanvir Hassan, S. M., Hassan, M., & Bharti, N. (2020). Urbanisation and water insecurity in the Hindu Kush Himalaya XE: Insights from Bangladesh, India, Nepal and Pakistan. *Water Policy, 22*(S1), 9–32. https://doi.org/10.2166/wp.2019.215

Tang, W., Zhou, T., Sun, J., Li, Y., & Li, W. (2017). Accelerated urban expansion in Lhasa City and the implications for sustainable development in a Plateau City. *Sustainability, 9*(9), 1499. https://doi.org/10.3390/su9091499

Thapa, R. B., & Murayama, Y. (2010). Drivers of urban growth in the Kathmandu valley, Nepal: Examining the efficacy of the analytic hierarchy process. *Applied Geography, 30*(1), 70–83.

Thayyen, R. J., Dimri, A. P., Kumar, P., & Agnihotri, G. (2013). Study of cloudburst and flash floods around Leh, India, during August 4–6, 2010. *Natural Hazards, 65*(3), 2175–2204. https://doi.org/10.1007/s11069-012-0464-2

Times of India. (2020, September 16). *Leh airport to expand into a world-class facility*. The Times Group. https://timesofindia.indiatimes.com/travel/travel-news/leh-airport-to-expand-into-a-world-class-facility/articleshow/78147709.cms

Tiwari, P. C., Tiwari, A., & Joshi, B. (2018). Urban growth in Himalaya: Understanding the process and options for sustainable development. *Journal of Urban and Regional Studies on Contemporary India, 4*(2), 15–27. https://doi.org/10.15027/45582

Tiwari, P. C., & Joshi, B. (2020). Challenges of urban growth in Himalaya with reference to climate change and disaster risk mitigation: A case of Nainital town in Kumaon Middle Himalaya, India. In A. P. Dimri, B. Bookhagen, M. Stoffel, & T. Yasunari (Eds.), *Himalayan weather and climate and their impact on the environment* (pp. 473–491). Springer International Publishing. https://doi.org/10.1007/978-3-030-29684-1_23

Town Planning Organisation. (2018). *Master plan of Kargil* (pp. 1–135). https://cdn.s3waas.gov.in/s341ae36ecb9b3eee609d05b90c14222fb/uploads/2018/09/2018091095-1.pdf

UN-HABITAT. (2010). *The state of Asian cities 2010/11*. United Nations Human Settlements Programme, Regional Office for Asia and the Pacific. https://unhabitat.org/the-state-of-asian-cities-201011

United Nations. (2019). *World urbanization prospects: The 2018 revision*. Department of Economic and Social Affairs, Population Division. https://population.un.org/wup/publications/Files/WUP2018-Report.pdf

Vaidya, R. A., Shrestha, M. S., Nasab, N., Gurung, D. R., Kozo, N., Pradhan, N. S., & Wasson, R. J. (2019). Disaster risk reduction and building resilience in the Hindu Kush Himalaya. In P. Wester, A. Mishra, A. Mukherji, & A. B. Shrestha (Eds.), *The Hindu Kush Himalaya assessment* (pp. 389–419). Springer International Publishing. https://doi.org/10.1007/978-3-319-92288-1_11

van Beek, M., & Pirie, F. (Eds.). (2008). *Modern Ladakh: Anthropological perspectives on continuity and change*. Brill.

Wang, Y., Wu, N., Kunze, C., Long, R., & Perlik, M. (2019). Drivers of change to mountain sustainability in the Hindu Kush Himalaya. In P. Wester, A. Mishra, A. Mukherji, & A. B. Shrestha (Eds.), *The Hindu Kush Himalaya assessment* (pp. 17–56). Springer International Publishing. https://doi.org/10.1007/978-3-319-92288-1_2

Wester, P., Mishra, A., Mukherji, A., & Shrestha, A. B. (Eds.). (2019). *The Hindu Kush Himalaya assessment: Mountains, climate change, sustainability and people*. Springer International Publishing. https://doi.org/10.1007/978-3-319-92288-1

World Bank. (2013). *Urbanization beyond municipal boundaries: Nurturing metropolitan economies and connecting peri-urban areas in India*. The World Bank. https://doi.org/10.1596/978-0-8213-9840-1

Zutshi, J. N., Farooqi, B. A., Din, G. M., & Nanda, J. K. (1961). *Census of India: Village survey monograph of Kharboo (Tehsil Kargil, District Ladakh), Part VI, No. 13, Vol. VI*. Director of Census Operations, Jammu & Kashmir.

Appendix A
Semi-structured Interviews

Water Supply

a. Can you describe the current state of water supply in Kargil town?
b. What challenges do you perceive in ensuring access to clean and reliable water supply for urban residents?
c. How has the water supply infrastructure evolved over time, and what improvements do you think are necessary?

Deep Drains

d. What are your thoughts on the deep drains system in Kargil town?
e. Have you noticed any issues with the deep drains, such as blockages?
f. Are there any specific areas in the town where problems with the deep drains are particularly prevalent?

Solid Waste Management

g. How would you describe the current situation regarding solid wastes management in Kargil town?
h. What are some of the challenges associated with solid waste collection, disposal, and recycling?
i. In your opinion, what kind of strategies could be implemented to improve solid waste management practices?

Public Toilets

j. Can you discuss the availability and condition of sanitation facilities such as public toilets?
k. What are some of the sanitation challenges faced by people in this town?
l. Are there any initiatives you believe could help address sanitation-related issues?

A. Hussain, *Urbanisation and Development Challenges in Kargil, Ladakh*, Advances in Asian Human-Environmental Research, https://doi.org/10.1007/978-3-032-05667-2

Urban Mobility

m. What are your thoughts on traffic congestion and vehicular traffic flow in this town?
n. How does congestion impact people's daily lives, mobility, and access to essential services?
o. What measures can be implemented to reduce congestion and improve traffic management?
p. How would you describe the condition of footpaths and pedestrian infrastructure?
q. Have you encountered any challenges or safety concerns while using footpaths?
r. What improvements do you think are needed to make footpaths more accessible, safe, and pedestrian friendly?

Other Interview Questions

s. Can you share your experiences of living in Kargil town and how you have observed its transformation over the years?
t. Can you provide specific examples of urban development schemes that have positively impacted the town population?
u. Are urban bylaws and regulations being effectively implemented in Kargil town? If so, how?
v. How do you see technological innovation influencing the future of urban development in Kargil?
w. What roles should different stakeholders, particularly LAHDC-Kargil, UT of Ladakh, and community organisations, play in addressing urban challenges and issues?
x. How can local residents contribute to the beautification of Kargil town?
y. What recommendations would you offer to policymakers to ensure sustainable and inclusive growth?

These questions were designed to gather participants' viewpoints, personal experiences, and insights regarding urban infrastructure challenges and related issues in Kargil town, Ladakh.

Appendix B
Occupational Structure and Village Population

See Tables B1, B2, B3, B4 and B5.

Table B1 Town-wise occupational distribution of workers in 1981

Towns	Main workers	Marginal workers	Main workers			
			Cultivators	Agricultural labourers	Household industry workers	Other workers
Kargil	1440	404	324	13	21	1082
Drass	161	223	24	0	0	137
Padum	90	0	85	1	0	4
Sankoo	688	152	469	39	1	179

Sources Census of India (1981)

Table B2 Town-wise occupational distribution of workers in 2011

Towns	Main workers	Marginal workers	Main workers			
			Cultivators	Agricultural labourers	Household industry workers	Other workers
Kargil	6102	1311	62	20	10	6010
Drass	6966	563	60	4	0	6902
Padum	249	39	4	0	14	231
Sanko	372	319	35	5	40	292

Sources Census of India (2011)

A. Hussain, *Urbanisation and Development Challenges in Kargil, Ladakh*, Advances in Asian Human-Environmental Research, https://doi.org/10.1007/978-3-032-05667-2

Table B3 Block-wise occupational distribution of workers in 2001

Blocks	Main workers	Marginal workers	Main workers			
			Cultivators	Agricultural labourers	Household industry workers	Other workers
Kargil	7930	2936	2195	31	140	5564
Drass	3136	2533	899	0	5	2232
Zanskar	4622	230	2476	7	9	2130
Sankoo	4505	2974	2925	2	27	1551

Sources Census of India (2001)

Table B4 Block-wise occupational distribution of workers in 2011

Blocks	Main workers	Marginal workers	Main workers			
			Cultivators	Agricultural labourers	Household industry workers	Other workers
Kargil	3378	3337	417	40	35	2886
Drass	8608	1933	109	16	60	8423
Zanskar	2113	3127	482	275	50	1306
Sankoo	1666	3550	158	26	67	1415

Sources Census of India (2011)

Table B5 Total population of the villages from 1981 to 2011

Villages	Total population in 1981	Total population in 2011	Villages	Total population in 1981	Total population in 2011
Abran	228	418	Nagmakausar	512	1126
Achambur	256	470	Namsuru	496	905
Akchamal	947	1909	Nunamchey	140	199
Akshu	361	332	Ofti pipiting	286	941
Apati	594	1245	Padum	139	1538
Ating	366	605	Pandrass	382	731
Barchey	317	555	Panikhar	156	325
Barsoo	927	1602	Parkachik	642	1138
Bartoo	348	679	Pashkyum	1609	3419
Batambis	184	332	Phey	332	509
Bhimbat	569	1393	Phoo	424	610
Bodhkharbu	704	970	Phukthal	91	280
Chah	487	465	Pipcha	103	213

(continued)

Table B5 (continued)

Villages	Total population in 1981	Total population in 2011	Villages	Total population in 1981	Total population in 2011
Chiktan	694	1285	Prantee	322	518
Chokiyal	398	965	Purtikchey	585	963
Choskor Suru	213	431	Ranbirpur (Drass)	599	8397
Choskore	1896	3430	Rangdum	173	368
Chulichan	443	912	Rantaksha	289	358
Chuliskambo	420	982	Remala Skaygam	201	588
Darchiks	305	624	Raru Money	161	329
Darkit	129	254	Salapi Ruruk	267	361
Faroona	442	893	Safi	478	639
Garkon	434	1287	Saleskote	1051	3605
Gindyal	244	676	Samrah	758	1441
Goshan	622	1485	Sangrah	657	1183
Gund Mangalpur	929	1894	Sani	854	614
Gyaling	158	271	Sanjak	271	416
Hagnis	886	1624	Salapigaipak	295	701
Hamiling	142	266	Shakar	1256	1922
Hardas	653	1480	Shanka	244	244
Heniskot	191	330	Shargol	258	486
Holiyal	308	615	Shargandi	141	283
Icher	559	249	Shilikchey	651	1006
Ichu	47	159	Shimsha	291	794
Jasgund	211	467	Silmo	851	1627
Kaksar	377	958	Skambo	317	623
Kanoor	681	1481	Stakpa	463	878
Karamba	201	397	Stakchey	877	1453
Karcheykhar	548	942	Tacha	260	467
Kargee	328	566	Taisuru	283	589
Kargyak	195	429	Tambis	1169	2442
Karkit	932	2373	Stongde Kumi	724	914
Karit	246	526	Thang Dumbur (Sankoo)	1389	2698
Karpokhar	85	133			

(continued)

Table B5 (continued)

Villages	Total population in 1981	Total population in 2011	Villages	Total population in 1981	Total population in 2011
Karshah	177	964	Tangol	364	694
Khachey	314	535	Techa Khasar	224	290
Khandi	748	1456	Testa	185	460
Kharbu	454	1074	Thasgam	167	400
Khows	377	756	Thasgam Thevena	1318	2567
Kochik	165	316	Thrangos	210	413
Kukshow	428	718	Thulspursa	197	420
Kukstey	187	349	Tingdo	180	292
Lalung	418	685	Tonrian Thagan	368	573
Lamsu Sandow	346	520	Toumel	459	1084
Lankarchey	1236	2425	Trespone	1424	3332
Lalung	418	685	Trankuchan (Tronjen)	431	1094
Latoo	91	280	Umba	520	844
Lotsum	217	405	Upti	941	1241
Lungmi Rezing	252	478	Wakha	1007	1773
Matayan	287	500	Youlboo	407	763
Minjee	1023	2331	Yogmakharbu	364	554
Mulbekh	915	1393	Yourbaltak	1107	2177
Muradbagh	332	691	Yuljuk	701	1196
Mushku	255	572	Zangla	345	954

Sources Census of India (1981, 2011)

Appendix C
Tourism and Vehicle Registrations

See Tables C1, C2 and C3.

Table C1 Total number of hotels, guest houses, and travel agencies in Kargil, Ladakh

Years	Hotels	Guest houses	Travel agencies
2015–16	17	70	40
2016–17	21	73	44
2017–18	34	80	57
2018–19	36	86	60
2019–20	42	87	66
2020–21	46	96	67
2021–22	47	110	76
2022–23	48	110	81
2023–24	49	115	87

Sources District Tourism Department of Kargil, UT of Ladakh

A. Hussain, *Urbanisation and Development Challenges in Kargil, Ladakh*, Advances in Asian Human-Environmental Research, https://doi.org/10.1007/978-3-032-05667-2

Table C2 International and domestic tourist arrivals in Kargil, Ladakh

Years	International	Domestic	Total
1976	4325	0	4325
1977	5866	0	5866
1978	8345	0	8345
1979	10165	0	10165
1980	10889	0	10889
1981	10992	0	10992
1982	11162	0	11162
1983	10385	0	10385
1984	8522	0	8522
1985	10301	0	10301
1986	11412	1855	13267
1987	10008	2165	12173
1988	11916	5495	17411
1989	11876	5625	17501
1990	924	210	1134
1991	1962	430	2392
1992	3148	430	3578
1993	3326	220	3546
1994	4792	780	5572
1995	1962	381	2343
1996	2255	740	2995
1997	3111	1020	4131
1998	2636	2230	4866
1999	2636	2230	4866
2000	190	132	322
2001	1510	1077	2587
2002	440	740	1180
2003	2044	2200	4244
2004	3271	3770	7041
2005	4289	8258	12547
2006	3460	9767	13227
2007	3240	7889	11129
2008	3648	13543	17191
2009	3179	16968	20147
2010	2984	25772	28756
2011	2692	54392	57084
2012	3155	57603	60758
2013	1634	46614	48248

(continued)

Table C2 (continued)

Years	International	Domestic	Total
2014	2204	47174	49378
2015	2393	59790	62183
2016	839	62168	63007
2017	4612	96886	101498
2018	6116	103442	109558
2019	3643	73476	77119
2020	0	0	0
2021	79	50750	50829
2022	2457	223086	225543
2023	4655	270310	274965
2024	4215	320432	324647

Sources District Tourism Department of Kargil, UT of Ladakh

Table C3 Total number of registered motor vehicles in Kargil, Ladakh

Years	Motor bicycles/ scooters	LGV/MGV/HGV	Buses/taxis/maxi cabs	LMV	Other vehicles
2014	341	417	1373	1691	478
2015	361	468	1499	1837	500
2016	361	546	1611	2040	517
2017	412	621	1681	2309	574
2018	509	730	1772	2680	629
2019	522	820	1824	3119	724
2020	531	949	1840	3441	825
2021	540	1097	1858	4074	914
2022	586	1268	1935	4547	1065

Sources Regional Transport Office Kargil of LAHDC-Kargil and MoRTHs, India, light motor vehicles (LMV), light goods vehicles (LGV), medium goods vehicles (MGV), and heavy goods vehicles (HGV)

Appendix D
Precipitation and Temperature

See Tables D1 and D2.

Table D1 Monthly mean precipitation, monthly mean temperature, and monthly mean maximum/minimum temperature in Kargil

Monthly mean (1981–2022)				
Months	Monthly mean precipitation	Monthly mean maximum temperature	Monthly mean minimum temperature	Monthly mean temperature
Jan.	27.9	−2.7	−27.8	−14.5
Feb.	30.5	−2.0	−27.1	−12.9
Mar.	36.1	3.3	−23.4	−8.0
Apr.	32.1	9.0	−13.6	−1.9
May	22.2	13.6	−6.8	2.8
Jun.	22.2	18.6	−2.8	7.0
Jul.	25.5	22.4	0.6	10.8
Aug.	26.7	22.0	0.0	10.5
Sep.	15.5	18.5	−4.4	6.0
Oct.	14.2	12.0	−10.2	−0.6
Nov.	18.7	5.4	−18.2	−6.3
Dec.	17.8	0.5	−23.2	−11.4

Sources NASA Langley Research Centre POWER Project funded through the NASA Earth Science Directorate Applied Science Program

A. Hussain, *Urbanisation and Development Challenges in Kargil, Ladakh*, Advances in Asian Human-Environmental Research, https://doi.org/10.1007/978-3-032-05667-2

Table D2 Total annual precipitation, annual mean temperature, and annual mean maximum/minimum temperature

Annual mean (1981–2022)				
Years	Annual mean maximum temperature	Annual mean minimum temperature	Annual mean temperature	Total annual precipitation
1981	22.2	−32.5	−1.9	337.5
1982	22.1	−34.6	−2.5	427.2
1983	24.3	−27.6	−2.4	358.6
1984	23.9	−33.1	−1.3	247.9
1985	23.0	−29.0	−0.9	258.4
1986	22.5	−34.7	−3.2	416.6
1987	21.4	−31.1	−2.3	263.7
1988	23.1	−26.5	−0.8	337.5
1989	24.1	−28.8	−2.4	284.8
1990	23.2	−29.4	−0.9	290.0
1991	22.7	−23.2	−0.7	100.2
1992	22.2	−28.8	−1.9	363.9
1993	21.9	−26.4	−1.2	195.1
1994	23.7	−32.4	−1.4	311.1
1995	22.4	−33.6	−2.7	369.1
1996	21.8	−32.2	−2.2	432.4
1997	25.2	−26.0	−1.4	168.8
1998	23.3	−22.0	0.2	105.5
1999	23.0	−28.4	−0.1	89.7
2000	23.5	−25.7	−0.6	84.4
2001	25.3	−23.1	−0.3	142.4
2002	24.2	−23.0	−0.5	147.7
2003	22.5	−30.2	−1.5	337.5
2004	22.2	−31.2	−1.2	237.3
2005	22.0	−32.0	−1.8	226.8
2006	24.5	−31.5	−1.4	379.7
2007	23.3	−28.4	−0.8	137.1
2008	22.7	−38.4	−1.9	284.8
2009	23.0	−30.5	−2.2	205.7
2010	22.6	−29.4	−0.8	342.8
2011	22.7	−31.0	−1.7	247.9
2012	23.7	−34.8	−2.0	258.4
2013	23.9	−30.0	−0.7	300.6
2014	24.3	−27.8	−1.7	353.3

(continued)

Table D2 (continued)

Annual mean (1981–2022)				
Years	Annual mean maximum temperature	Annual mean minimum temperature	Annual mean temperature	Total annual precipitation
2015	22.4	−28.6	−1.1	532.6
2016	21.4	−26.5	0.1	237.3
2017	22.0	−31.1	−1.1	400.8
2018	23.7	−27.1	−1.3	337.5
2019	21.9	−33.5	−3.0	522.1
2020	21.8	−33.1	−2.8	421.9
2021	24.9	−31.8	−2.1	438.2
2022	22.6	−32.0	−1.2	475.8

Sources NASA Langley Research Centre POWER Project funded through the NASA Earth Science Directorate Applied Science Program

Appendix E
Accuracy Assessment

See Tables E1, E2, E3 and E4.

Table E1 Accuracy assessment of the LULC of Kargil from 1965 to 2020

Class	Ground truth (%) in 1965				CE (%)	OE (%)	PA (%)	UA (%)
	BU	ILV	BA	Total				
BU	21.4	0	0	7.32	0	78.6	21.4	100
ILV	45.3	84.4	0	37.4	41.3	15.6	84.4	58.7
BA	33.3	15.6	100	55.28	27.9	0	100	72.1
Total	100	100	100	100	OA = 69.1%; KC = 0.53			
Class	Ground truth (%) in 1979				CE (%)	OE (%)	PA (%)	UA (%)
	BU	ILV	BA	Total				
BU	33.3	0	0	10.2	0	66.67	33.3	100
ILV	28.8	91.7	0	34.3	25.7	8.33	91.7	74.3
BA	37.9	8.3	100	55.5	25	0	100	75
Total	100	100	100	100	OA = 77.3%; KC = 0.65			
Class	Ground truth (%) in 2000				CE (%)	OE (%)	PA (%)	UA (%)
	BU	ILV	BA	Total				
BU	79.5	0	1	24.2	1.5	20.5	79.5	98.5
ILV	11.2	100	0.2	36.9	9.3	0	100	90.8
BA	9.3	0	98.8	38.9	7.1	1.2	98.8	92.9
Total	100	100	100	100	OA = 93.5%; KC = 0.90			

(continued)

A. Hussain, *Urbanisation and Development Challenges in Kargil, Ladakh*, Advances in Asian Human-Environmental Research, https://doi.org/10.1007/978-3-032-05667-2

Table E1 (continued)

Class	Ground truth (%) in 2020				CE (%)	OE (%)	PA (%)	UA (%)
	BU	ILV	BA	Total				
BU	92.3	0	0	36.1	0	7.7	92.3	100
ILV	7.7	100	0	36.1	8.4	0	100	91.6
BA	0	0	100	27.8	0	0	100	100
Total	100	100	100	100	OA = 96.9%; KC = 0.95			

Sources Satellite Imagery—Corona (8 Oct 1965), Corona (17 June 1979), ASTER (4 September 2000), and PlanetScope (8 June 2020); Built-up area (BU), Irrigated land and vegetation (ILV), Barren area (BA), Commission error (CE), Omission error (OE), Producer's accuracy (PA), User's accuracy (UA), Overall accuracy (OA), and Kappa coefficient (KC)

Table E2 Accuracy assessment of the LULC of Padum from 1980 to 2021

Class	Ground truth (%) in 1980				CE (%)	OE (%)	PA (%)	UA (%)
	BU	ILV	BA	Total				
BU	31.9	0	0	10.1	0	68.1	31.9	100
ILV	52	93.2	0	46.2	35.6	6.8	93.2	64.4
BA	16.1	6.8	100	43.7	16.7	0	100	83.3
Total	100	100	100	100	OA = 76.3%; KC = 0.64			
Class	Ground truth (%) in 2005				CE (%)	OE (%)	PA (%)	UA (%)
	BU	ILV	BA	Total				
BU	33.3	0	0	10.3	0	66.7	33.3	100
ILV	22.2	100	0	43.1	16	0	100	84
BA	44.4	0	100	46.6	29.6	0	100	70.4
Total	100	100	100	100	OA = 79.3%; KC = 0.68			
Class	Ground truth (%) in 2021				CE (%)	OE (%)	PA (%)	UA (%)
	BU	ILV	BA	Total				
BU	96.2	9.9	0	25.2	13.3	3.8	96.2	86.7
ILV	0	90.1	0	30.7	0	9.9	90.1	100
BA	3.8	0	100	44.1	2	0	100	98
Total	100	100	100	100	OA = 95.8%; KC = 0.93			

Sources Satellite Imagery—Corona (11 September 1980), ASTER (13 October 2005), and PlanetScope (24 July 2021); Built-up area (BU), Irrigated land and vegetation (ILV), Barren area (BA), Commission error (CE), Omission error (OE), Producer's accuracy (PA), User's accuracy (UA), Overall accuracy (OA), and Kappa coefficient (KC)

Table E3 Accuracy assessment of the LULC of Sankoo from 1971 to 2021

Class	Ground truth (%) in 1971				CE (%)	OE (%)	PA (%)	UA (%)
	BU	ILV	BA	Total				
BU	15.8	0	0	3	0	84.2	15.8	100
ILV	75.4	88.2	14.7	56	36.3	11.8	88.2	63.7
BA	8.8	11.8	85.3	41	15.7	14.7	85.4	84.3
Total	100	100	100	100	OA = 73.2%; KC = 0.55			
Class	Ground truth (%) in 2000				CE (%)	OE (%)	PA (%)	UA (%)
	BU	ILV	BA	Total				
BU	57.1	0	0	14	0	42.9	57.1	100
ILV	42.9	96.1	17.7	59.7	26.5	3.9	96.2	73.5
BA	0	3.9	82.3	26.3	6.7	17.7	82.4	93.3
Total	100	100	100	100	OA = 82.4%; KC = 0.71			
Class	Ground truth (%) in 2021				CE (%)	OE (%)	PA (%)	UA (%)
	BU	ILV	BA	Total				
BU	78.1	0	0	24.6	0	21.9	78.1	100
ILV	21.9	99.6	3.4	42.8	18.8	0.4	99.6	81.2
BA	0	0.4	96.6	32.5	0.4	3.4	96.6	99.6
Total	100	100	100	100	OA = 91.8%; KC = 0.87			

Sources Satellite Imagery—Corona (28 September 1971), ASTER (18 July 2000), and PlanetScope (12 August 2021); Built-up area (BU), Irrigated land and vegetation (ILV), Barren area (BA), Commission error (CE), Omission error (OE), Producer's accuracy (PA), User's accuracy (UA), Overall accuracy (OA), and Kappa coefficient (KC)

Table E4 Accuracy assessment of the LULC of Drass from 1965 to 2021

Class	Ground truth (%) in 1965				CE (%)	OE (%)	PA (%)	UA (%)
	BU	ILV	BA	Total				
BU	90.9	0	0	25.3	0.0	9.2	90.9	100
ILV	6.3	100	28.2	42	34	0	100	66
BA	2.8	0	71.8	32.7	2.4	28.2	71.8	97.6
Total	100	100	100	100	OA = 84.9%; KC = 0.77			
Class	Ground truth (%) in 2000				CE (%)	OE (%)	PA (%)	UA (%)
	BU	ILV	BA	Total				
BU	37.8	0	0	11.7	0	62.2	37.8	100
ILV	46	100	0	48.3	29.3	0	100	70.7
BA	16.2	0	100	40	12.5	0	100	87.5
Total	100	100	100	100	OA = 80.8%; KC = 0.71			

(continued)

Table E4 (continued)

Class	Ground truth (%) in 2021				CE (%)	OE (%)	PA (%)	UA (%)
	BU	ILV	BA	Total				
BU	94.6	0	0	36.5	0	5.4	94.6	100
ILV	5.4	100	0	43.8	4.8	0	100	95.2
BA	0	0	100	19.8	0	0	100	100
Total	100	100	100	100	OA = 97.9%; KC = 0.96			

Sources Satellite Imagery—Corona (8 October 1965), ASTER (18 July 2000), and PlanetScope (18 July 2021); Built-up area (BU), Irrigated land and vegetation (ILV), Barren area (BA), Commission error (CE), Omission error (OE), Producer's accuracy (PA), User's accuracy (UA), Overall accuracy (OA), and Kappa coefficient (KC)

Glossary

Purgi/Balti English

Bal Wool
Brangsa Halting place
Bras Rice
Cha Tea
Changpas The nomadic pastoralists people inhabiting in the Changthang region
Chapa The annual trade caravan from Lhasa to Ladakh was known as Chapa (consisting of 200 animal loads of tea)
Charas Cannabis
Charpha Rainfall
Chugzam Suspension bridge
Chuli Apricot
Dzo Offspring of the male yak and the common cow
Gompa A Tibetan Buddhist monastery
Gyalpo King
Jhing Cultivated land or cultivated field
Kah Snow
Kalaq Clay
Khachul Kashmir
Khar Castle, fort
Khara Sugar
Khatak Ceremonial scarf
Kro Wheat
Kushu Apple
La Mountain pass
Lopchak The regular trade/tribute mission from Leh (Ladakh) to Lhasa every three years with offerings to the Dalai Lama at the Monlam Prayer Festival
Lut Manure
Malchang Willow

A. Hussain, *Urbanisation and Development Challenges in Kargil, Ladakh*, Advances in Asian Human-Environmental Research, https://doi.org/10.1007/978-3-032-05667-2

Nas Barley
Ol Meadow, lucerne
Owse Mulberry
Pagbu Brick
Phey Flour (barley flour—nasphey; wheat flour—baqphey)
Rzing Man-made pond or reservoir
Shingzam Wooden bridge
Tsonglen Trade
Yourba Water channel
Yul Village, country
Zamba Bridge
Zberpa or yoolat Poplar
Urdu English
Bazaar A marketplace
Kiraiyakash Transporters for the goods who would lead livestock along the route or trade caravans
Serai An inn or a place of accommodation for travellers and traders
Wazir-i-Wazarat A governor or government official for the Maharaja's government of Jammu and Kashmir
Road Name

NH-1 Srinagar-Kargil-Leh highway
NH-3 Manali-Leh highway
NH-301 Kargil-Zanskar highway
NPD Nimmu-Padum-Darcha road

Index

A. Hussain, *Urbanisation and Development Challenges in Kargil, Ladakh*, Advances in Asian Human-Environmental Research, https://doi.org/10.1007/978-3-032-05667-2

MIX
Papier aus verantwortungsvollen Quellen
Paper from responsible sources
FSC® C105338

FSC
www.fsc.org

If you have any concerns about our products,
you can contact us on
ProductSafety@springernature.com

In case Publisher is established outside the EU,
the EU authorized representative is:
Springer Nature Customer Service Center GmbH
Europaplatz 3, 69115 Heidelberg, Germany

Printed by Libri Plureos GmbH
in Hamburg, Germany